This book and your GCSE course

Board	Specifications available	Full course	Short course
AQA www.aqa.org.uk	Electronic Products	3541	3551
	Food Technology*	3542	3552
	Graphic Products*	3543	3553
	Product Design	3544**	N/A
	Resistant Materials Technology*	3545	3555
	Systems & Control Technology	3546	N/A
	Textiles Technology*	3547	3557
EDEXCEL www.edexcel.org.uk	Food Technology	1970	3970
	Graphic Products	1972	3972
	Resistant Materials Technology	1973	3973
	Systems & Control Technology	1974***	3974***
	Textiles Technology	1971	3971
OCR www.ocr.org.uk	Electronic Products	1953	1053
	Food Technology	1954	1054
	Graphic Products	1955	1055
	Industrial Technology	1959	N/A
	Resistant Materials Technology	1956	1056
	Systems & Control Technology	1957****	N/A
	Textiles Technology	1958	1058
WJEC www.wjec.org.uk	Food Technology	0139	0270
	Graphic Products	0140	0271
	Resistant Materials Technology	0141	0272
	Systems & Control Technology	0142	0273
	Textiles Technology	0143	0274
	Industrial Technology	0144	0275
NICCEA www.ccea.org.uk	Design & Technology: Product Design		
	Home Economics		
	Technology & Design		

Visit your awarding body website for full details of your course or download your complete GCSE specifications.

Use these pages to get to know your course
- Make sure you know your examination board
- Know how your course is assessed
- How many papers and how long?
- How much coursework?

Coursework	Examination
Coursework 60% Full course – 40 hours Short course – 20 hours	Written paper – 40% Full course – 2 hours Short course – 1.5 hours * Preparation sheet issued ** Paper A – Core + Mechanisms Paper B – Core + Pneumatics
Coursework 60% Full course – 40 hours Short course – 20 hours *** Project can be: electronic, mechanical or electro-mechanical	Written paper – 40% Full course – 1.5 hours Short course – 1 hour *** Electronics or Mechanisms
Coursework 60% Full course – 40 hours Short course – 20 hours	Written paper – 40% Full course – (H) 2 × 1.25 hours, (F) 2 × 1 hour Short course – (H) 1.25 hours, (F) 1 hour except Graphic Products: + 15 minutes on each paper. **** Core + Electronics or Pneumatics or Mechanisms
Coursework 60% From 1 or more focus areas	Written paper – 40% Full course – 2 hours Section A: Core – 30 minutes Section B: Focus – 1.5 hours Short course – 1 hour Core + design question on Focus

Preparing for the examination

Once all your coursework is finished, your teachers will either be completing areas of the specification not already covered at this stage, revising the work already covered in order to prepare you for the terminal examination, or practising questions under controlled conditions.

Do not rely on this alone, as you could have **specific needs** that your teacher may not cover when preparing work for the group as a whole.

Start revising **as early as possible**, especially once time-consuming elements like coursework have finished. The main purpose is to find any gaps in your knowledge and ask the necessary questions before it is too late.

Preparing a revision plan

- Actually **write down** the plan on paper, noting all the critical dates and times on it, e.g.:
    - all your examination dates
    - remaining coursework deadlines
    - study leave
    - family occasions such as birthdays, holidays, etc
    - personal events such as sporting fixtures, days out, etc
    - allow some time for relaxation and exercise – you need to be 'fit' for examinations!
- Divide the rest of the time between **all of your subjects**. When allocating time to each subject, take into account such factors as:
    - how well you did in the mock examinations
    - how easy you generally find the subject
    - how important you consider the subject is with respect to your future plans
    - the actual date of each examination – working back from the last one
    - your preferred method of revising. Don't allocate yourself time that you will not use effectively. Variety created by 'little and often' or 'topic by topic' may be more effective
    - you could use colour-coded blocks for each subject, to give you a clearer indication of how you have allocated time to each subject.
- Don't be afraid to alter your plans as circumstances change, but try to **stick to it** as closely as possible!

Revision methods

Revision requires your notes, textbooks, other aids such as this revision guide, **time,** application and practice questions (often in the form of past papers and mark schemes). Taking into account the following could increase the effectiveness of your revision:

- Do not just read your notes or a textbook. You should **highlight points** as you go, using **'indexing tabs'** to locate important aspects and creating **'crib cards'** for later, quicker, run throughs.
- Sometimes work with a **friend or willing relative**. They can test you as you go and prompt you where necessary.
- Practise questions under **strict time conditions**, as many examinations can turn into 'time trials' if you are not used to working at the required speed.
- Revise in **comfortable but suitable** surroundings. Whereas music may help, a television is more likely to distract.
- Be **organised and tidy**. Time wasted searching for material in a pile on the floor or on your desk can never be recovered!
- Practise examination skills as well. Some exams last over two hours, so you must be able to think and, more importantly, **write neatly** for that long. This will be especially important for practical skills such as sketching, drawing and rendering.

Examination advice to boost your grade

Reading the question paper

- **Read the instruction rubric** well. It contains important information such as the length of the paper, number of questions to be attempted, existence of information sheets, and so on.
- **Read the whole paper** before you select your questions (if you have a choice), making sure you that you definitely attempt the questions that allow you to gain the most marks possible.
- Make sure you **read each question** very carefully and answer only what you have been asked.
- Note that **highlighted words in bold type** are a further reminder of what is required.
- Look for **clues** given by the Examiner. Sometimes answers are given in the text of the question, or can be found in other parts of the paper.
- If you have a choice, don't be fooled into choosing a question with an apparently easy start. Remember that the early parts to a long question are often easier than the later parts, which may carry more marks.

Answer the questions asked

- **Allocate time** to each question in proportion to the marks available. Be strict with yourself – it is pointless to lose the chance of answering a ten-mark question by spending too much time on a five-mark question! Remember, the marks available will be indicated in brackets at the end of each part of a question.
- Be careful not to rush your answers. An illegible answer cannot gain any marks, neither can a correct answer to a question that has not been asked!

Check your answers

- Plan to leave yourself time to **reread** each response and, if you finish early, never sit there doing nothing.
- Raise your hand if you have any problems, but be aware that invigilators may not answer all queries.
- **Keep calm!** Use the time available to your best advantage.
- Above all, **never give up** or panic.

Examination content

- Chapters 1, 2 and 8 are common to all specifications and should be studied in depth.
- There is a wide range of specifications offered by the Examination Boards, with overlapping content depending on the area of focus. The table below gives you an indication of the chapters that you should study all (A) or part of (P), in order to prepare for the examination.

Specification (area of focus)	Chapter 1	Chapter 2	Chapter 3	Chapter 4	Chapter 5	Chapter 6	Chapter 7	Chapter 8
Electronic products	A	A		P	P	P		A
Food technology	A	A	A	P				A
Graphic products	A	A		A	P			A
Industrial technology	A	A		P	P	P		A
Product design	A	A	P	P	P	P	P	A
Resistant materials technology	A	A		P	A			A
Systems and control technology	A	A		P	P	A		A
Textiles technology	A	A		P			A	A

Common assessment elements

Assessment objectives

The three English Examination Boards – **AQA, EDEXCEL and OCR** – all share the same assessment objectives.

- AO1   Classification and selection of materials and components
    Preparing, processing and finishing materials
    Manufacturing commercial products (industrial practices)
- AO2   Designing and making quality products in quantity
- AO3   Evaluating processes and products
    Design and market influences on society

Whilst covering the same basic assessment objectives, the Boards for Wales and Northern Ireland split them up differently.

WJEC includes:

- Developing, planning and communicating ideas
- Working with tools, equipment, materials and components to produce quality products
- Evaluating processes and products
- Knowledge and understanding of materials and components
- Knowledge and understanding of systems and control

NICCEA examines:

- Designing
- Communicating
- Manufacturing
- Using energy and control

Depending on your chosen **material specification**, you will cover these assessment objectives in different ways. However, there is a body of knowledge and understanding that is **common to all the specifications**, and this is covered in this chapter.

KEY POINT Only the applications will be different, and you must look for specific examples as evidence of your own personal knowledge and understanding through your coursework and examination responses.

Designing skills

Central to your success within Design and Technology is your ability to design and **produce solutions** to **identified problems**. Because designing is **not** always a **linear process**, it is sometimes difficult to have a clear understanding of the whole design process. A useful model is shown in Fig 1.1, where the **evaluation loops** or 'feedback' are clearly shown.

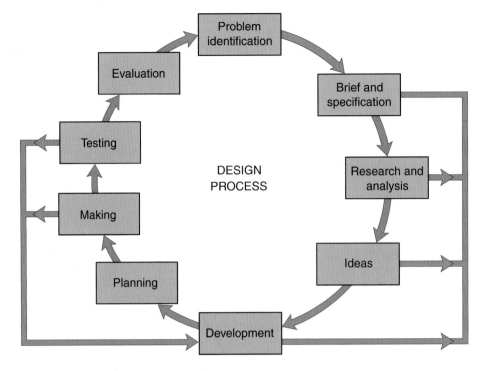

See Chapter 2, Coursework for more information about the design process.

Fig 1.1 A design process model

 KEY POINT Although you would normally start at the top with problem identification, it is also possible to start elsewhere, e.g. evaluating an existing product may lead to a further design problem to be solved.

Materials and components

Chapter 2, Coursework, is common to all material specifications, but you will find some of the content of other specifications useful – especially Chapter 4, Graphic products.

This objective is fully covered in **Chapters 3–7**, depending on the material specification you are studying. Although there are no chapters dedicated to Product Design or Industrial Technology, the majority of the content is covered within the other chapters.

Industrial practices

The following terms are commonly used with respect to manufacturing products **commercially**, and you should not only be able to respond using relevant examples in the examination, but you should show evidence of considering their implications in your coursework.

Quality control

This is the process of **checking for accuracy** throughout the manufacturing process, e.g. size, weight, colour, consistency, fit, etc.

Quality assurance

At various stages in the process it will be necessary to **build in procedures** to ensure that a certain quality is achieved, e.g. setting the depth stop on a drill, sensory analysis tests of food products, colour control strips in printing.

Tolerance

No manufacturing process can produce identical products. The **limits within** which the product **must fall** are known as the **tolerance** and are usually expressed as + **or** – (plus or minus). This can be applied to any aspect of the specification, e.g. size, weight and colour.

> **KEY POINT**
> It is important to realise that in the case of two components that must fit together, the cumulative effects of the tolerances must also be taken into account.

Relating your design to these production methods is an important aspect of industrial pratices.

Manufacturing in quantity

You will need to have an understanding of the **relative scales of production** possible for your product and the effect that these may have on such things as the materials used and the cost of the product.

- **One-off** (job production) refers to the making of one, or very few, hand-made/crafted items. They are usually **relatively expensive** because of the need to recover all of the costs of design and manufacture from the small number produced. They usually require **skilled workers** and can use expensive materials. Examples could include hand-made furniture or a wedding dress.

> **KEY POINT**
> **Prototypes** are made this way, which partly explains the high cost of product development.

- **Batch production** (batch flow) is a method of **low-level production**, say 500 units. This often uses the same basic equipment, materials and processes, but changes the final product to suit a **particular order** or market circumstance. Jigs and other time-saving devices are often made to aid production, and stored for possible future use. Examples could include personalised stationery or souvenirs of an event.

- **Repetitive batch** (flow) is used where batches are **regularly repeated**, but are not large enough to dedicate a production line fully to the product. Examples of this are **seasonal goods**.

- **Continuous production** (flow) is only used where **1000s** of the product are required **continually**. The initial set-up costs are usually **high**, often requiring **specialist** equipment and machines, and are **less reliant** on skilled workers. **Computer-aided manufacturing** (CAM) is also often used. Examples could include volume car production and bread.

> **KEY POINT**
> A commonly used term for flow production is mass production. The main factor here is that the fixed costs are spread over a large number of the product, thus reducing the unit cost.

- **Just-in-time production** reduces the **costs** of storage, unnecessary production, changes in market conditions, etc. Products are made **specifically to order** in terms of numbers and delivery time, using specially ordered quantities of materials and pre-booked use of machinery and plant. This method requires effective **planning** regimes and good supply/delivery chains. Examples could include components for electrical goods with continuously changing specifications such as personal computers.

Systems and control

KEY POINT A **system** is a number of parts, processes or actions that work together to perform a **function**.

KEY POINT **Control** involves one device having a direct effect on the action of an independent device **connected to it**.

The three elements of any system are: **INPUT**, **CONTROL/PROCESS**, **OUTPUT**.

An **input** is usually some kind of movement, e.g. pressing a switch, turning a key or a change in the environment such as a temperature rise, change in volume or light level.

> You should refer to these control elements in your coursework when using various pieces of equipment.

The **control** (or **process**) may be the change in size of the input, e.g. an amplifier increasing the level of sound, or the changing of the input into a different form, e.g. a rise in temperature opening a window vent in a greenhouse.

The **output** could be in the form of movement, sound, light or heat.

KEY POINT This simple form of control system is called an **open loop** system.

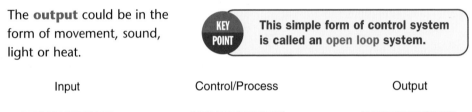

| Input | Control/Process | Output |

Fig 1.2 The open loop control system

The problem with open loop systems is that they **do not give any control over the output** other than that pre-set at the time of manufacture. For example, once the hi-fi system in Fig 1.3 has been turned on it will stay at the same volume.

Input   Control/Process   Output

CD player → Amplifier → Loudspeakers

Fig 1.3 An open loop hi-fi control system

To turn the sound up or down, a volume control is required. This is called **feedback**.

KEY POINT A control system that uses **feedback** is called a **closed loop** system.

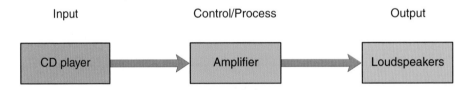

Input   Control/Process   Output

CD player → Amplifier → Loudspeakers

Volume control

Feedback

Fig 1.4 A closed loop hi-fi control system

However, the feedback in the hi-fi system in Fig 1.4 is provided **manually**. More usually, the feedback will be set to suit the user's requirements, or be capable of being so.

An example of an adjustable, automatic closed loop system is a central heating system. Householders require the house to be warm at fixed times and so a programmable timer is provided. They also require the temperature to remain reasonably constant at a comfortable temperature of their choosing and so a temperature sensor (or room thermostat) is also included.

> The decisions that control systems have to take can be shown as a logic diagram or truth table. A full explanation of logic circuits can be found in Chapter 6, Systems and control technology.

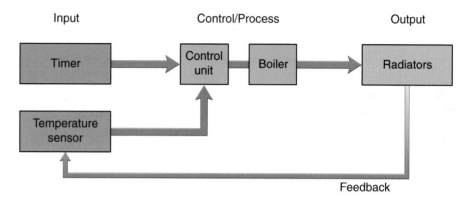

Fig 1.5 An automatic closed loop control system for a central heating unit

Computer-aided design and manufacture

Computer-aided design (CAD)

> You should investigate the specific uses of CAD in your particular material area of study.

CAD systems are widely used in industry even though they can be expensive to install and update. However, as more and more applications are developed, it becomes more important for businesses to take advantage of the possibilities that the systems offer. This enables them to remain **competitive** because of the **potential savings** that can be made.

> **KEY POINT**
> There are many different types of CAD program available, concerned with both the visual appearance of the product and the technical elements.

Types of 3D modelling
- **Wire-frame modelling** – the object is represented by a series of significant contour lines, which create a form of mesh
- **Surface modelling** – the surface finish can be added to produce a more realistic 3D image
- **Solid modelling** – the physical properties such as mass and volume can be determined

Computer simulations
Once the product has been modelled, CAD programs can be used to:

- **test** the product by analysing and predicting working stress
- provide a **virtual prototype** so that the product can be visualised from different angles and in different working situations

Use of databases
The development of some products can be assisted by the use of **databases**. Examples include using stored standard components for electronic circuits, calculating the nutritional value of food products and assessing the material costs of a product.

Advantages of using CAD

- The process is generally **quicker**, especially because **changes** can be easily made.
- The **quality** can be improved because of accuracy and information gained from simulations.
- The designs can be easily **stored, retrieved and communicated**.

Computer-aided manufacture (CAM)

CAM is where parts of the manufacturing process are carried out by equipment controlled by computer.

You should investigate the specific uses of CAM in your particular material area of study.

Computer numerical control (CNC)

CNC machines can be programmed to carry out tasks precisely and repetitively. These **automated** machines can also be used for transporting components, tools and materials around **a computer-controlled production line**.

The common term for these machines is robots.

Computer integrated manufacture (CIM)

CIM is the further development of the process and links CAD and CAM to produce a **fully automated** production system.

Advantages of using CAM

- It improves the **productivity** of batch production because the resetting of the machine for the different batches can be done very quickly, as the information is stored **electronically**.
- The machines can work **continuously** as long as they are maintained properly.
- There is greater **consistency** of outcome, with fewer faulty items as long as the quality assurance procedures are in place.
- It can readily be used with **hazardous** materials and processes.

Quality

To be successful in Design and Technology, you must demonstrate quality in your designing and making. You must also show an appreciation of what quality is in **commercial products**. The criteria will include:

- **Aesthetics**
  - Sensory qualities including touch, smell and taste
  - Visual qualities such as shape, form, balance, symmetry, proportion, rhythm – combining to produce a pleasing product

- **Function**
  - Performance compared with the specification
  - Consistency

A product investigation and disassembly are effective ways of assessing quality and health and safety issues.

- **Materials**
  - Sympathetic use in terms of such things as strength, weight, aesthetics and cost

- **Accuracy**
  - Precision and reliability

Evaluation

 KEY POINT **It is important to recognise that evaluation is an integral part of designing, and that it will form a major element of the examination and coursework.**

You must be able to:

- formulate **evaluation criteria** including a specification check
- devise ways of **testing the product** by means of questionnaires, focus groups, performance testing
- identify ways of including the **views of experts** in your evaluation

Health and safety

 KEY POINT **You should not only include evidence in your coursework that you have worked in a safe manner, but that you understand the effect that Health and Safety legislation has in industry.**

You should be able to:

- assess **hazards** and **risks**, both in the manufacture and the use of a product, taking steps to control the risks
- recognise and use the information available regarding legislation aimed at protecting the public
- understand the significance of the symbols/signs related to quality assurance

Examples of the issues that could be considered are:

- child- and tamper-proofing of packaging
- inherent dangers with electrical goods
- fire-retardant materials for textile products
- hygiene regulations concerned with the storage and preparation of food products
- safe working practices when using hand and machine tools

Design and market influences on society

 KEY POINT **Throughout the design process, whether in your coursework or the examination, you must take account of the social, cultural, moral and environmental issues involved.**

Depending on your material area of study, this could include:

- not offending **minority groups**
- considering wider **cultural influences** as the **basis** for design
- taking account of the environmental effects of the product in terms of its **production, use and disposal**
- understanding the **historical context** of development and its **effect on society**
- being able to make **moral judgements** on the **consequences** involved in designing

Coursework

General requirements

Always keep your projects safely and neatly stored, including any rough work and development models.

Your **Key Stage 3** work in Design and Technology will have given you the building blocks necessary to prepare you for this element. These skills will be enhanced and extended during the first year of Key Stage 4 by a number of **minor projects**. It is important that you put your full effort into these pieces of work, as there are certain circumstances when they may be used for assessment purposes e.g. prolonged absence in your final year due to injury or illness.

Throughout the project folio you should address the following issues arising from your work:

- Industrial and commercial practices
- Spiritual
- Moral
- Ethical
- Social

- Cultural
- Environmental
- Health and safety
- European

In addition you will need to include evidence of the **appropriate use** of:

- ICT (Information and Communication Technology)

This is the major component of the examination – usually 60%. However, it is also the element over which you have complete control. Use your time wisely and efficiently.

KEY POINT

Remember that you and your teacher will have to sign that the work presented for assessment is all your own. It is worth noting that any work submitted that cannot be authenticated by your teacher, or is considered unlikely to be by you, may be removed from your submission. In extreme circumstances, the Examination Board may take further action that could jeopardise your other examination entries.

Quality of written communication

Throughout your coursework design folder, you will also be assessed on your ability to:

- present **relevant information** in a form that suits its purpose
- ensure that text is legible and that **spelling, punctuation and grammar** are accurate, so that the meaning is clear

Choosing a project for success

Experience has confirmed that students who **devise their own project outlines** generate the best projects, because they are likely to be more highly motivated. However, most Examination Boards publish **lists of suitable project outlines** to act as a starting point for most candidates.

It is advisable for all students to study these, along with the assessment criteria, as a guide to the type of work and the standards required.

Your project outline should allow you to:

- show an **integrated approach** to designing and making
- work at a level that demonstrates your full ability
- cover all the assessment objectives

It should also offer you the scope to consider:

- the effects of the technological activities involved
- systems and control
- methods of production
- maintenance

- CAD/CAM
- quality assurance
- health and safety

 KEY POINT **Choose a project that interests you and that you can sustain for the length of time available.**

> The best projects tend to be 'real' ones, generated by a need that someone has brought to your notice. This has the added advantage of having someone else to act as a consultant, help generate the specification, test the outcome and provide effective evaluation.

Do:

- choose a project that will allow you to work mainly in the medium which best suits your **skills**
- check that you have access to all the **resources** required to carry out the project
- ensure that you can complete your project in the available time, given your other subject commitments

Don't:

- start with the outcome. '**I want to make a . . .**' is an **unpromising** beginning.
- choose what your friend has decided to do. He/she may have a real interest or expertise that you don't possess.

Group projects

The most common form of project is one that is carried out by an individual working alone. In some circumstances it may be appropriate to work as part of a team on a project that is made up of a number of smaller elements that together create a much larger outcome.

If you choose to embark on such a project you must ensure that:

- your contribution to the outcome is **clearly identifiable**
- you have **covered all aspects** of the assessment objectives and that you are not relying on another student to carry out any part of it

KEY POINT **Remember that if you become involved with a group project, you may not own the end result, and therefore will have only photographs as evidence in the future!**

Managing the project

It is essential that you cover all aspects of the assessment objectives. Therefore you must **avoid** spending too much time on any one aspect of the project, however well it is done, as it will not compensate fully for rushed, incomplete or missing elements.

It is advisable to:

- produce a **time chart** (see Fig 2.1) for the whole of the project at the beginning. Include such things as:
  - holidays
  - mock examinations
  - school events
  - other subject deadlines
  - family occasions

- allow **slack time** for such things as:
  - awaiting replies to letters
  - arranging visits and interviews
  - developing photographs
  - access to specialist facilities
  - ordering materials
- allocate the most suitable **presentation methods** to the various parts of the project in order to present an appropriately wide range of skills, **including ICT**

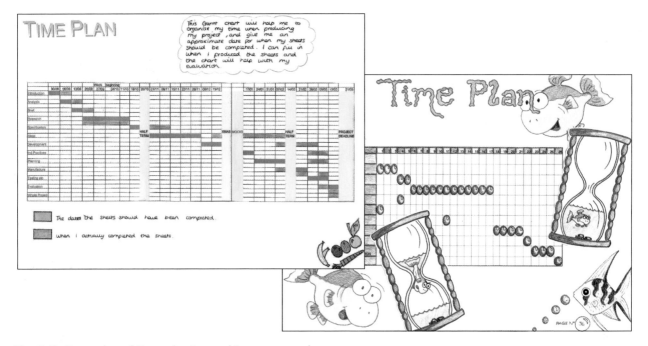

Fig 2.1 Examples of time charts used in coursework

> Most examination boards accept notebooks which will often gain marks by providing additional evidence that you may have neglected in your main folder.

- maintain a **project notebook**. This should be a true account of your progress throughout the entire project and should include:
  - a diary of events
  - thoughts and conversations
  - addresses, telephone and fax numbers
  - internet addresses of relevant sites
  - sketches
  - cuttings, etc
  - important things to remember or do at a later stage

Folder format

Although marks are mainly gained from the **content of the folder**, there are always some marks available for '**Graphical communication and the use of ICT**'. If time allows, consider including some or all of the following:

- A front cover containing:
  - your name and candidate number
  - your school's name and number
  - a title and relevant illustration
- A contents page
- A time plan
- An appendix containing such things as reports, questionnaires, etc that are referred to in the main body of the folder
- A bibliography/acknowledgements page containing:
  - sources of information referred to
  - ICT programs used
  - help received from other people

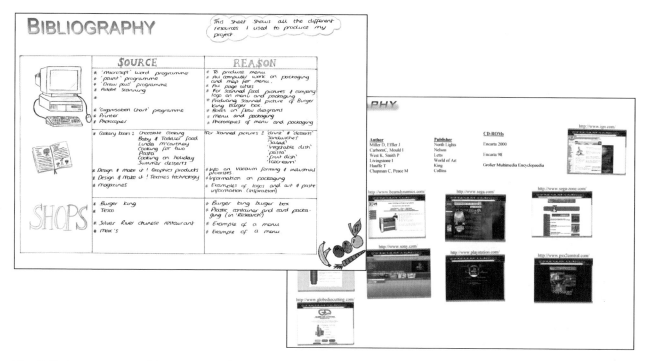

Fig 2.2 Resources used in coursework

Presentation methods

Page layout

It is an advantage if you can **personalise** your work and create a **style of presentation** that makes it easy to follow. These guidelines may be helpful.

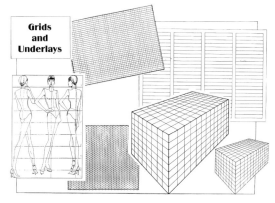

- Keep a **consistent** style throughout.
- **Simple** borders are more effective.
- Title blocks should **not dominate** the page.
- **Limit** the range of colours used on a page.
- Number pages **after completing** the project so you don't use 'a' numbers.
- Plan the use of the space available to **avoid overcrowding or large blank areas** – relevant illustrations/photographs/cut and paste are useful here.
- Use **underlays** with grids and lines to align your work.

Fig 2.3 Examples of underlays used to assist page layout

> Remember that a fully annotated sketch can often be more appropriate than a large passage of text.

The written word

It will be necessary within your folder to include a substantial amount of writing. The main consideration is to be **concise**. Only include material that is relevant to the progress of the project.

It is here that you can also use ICT to your advantage because it:

- generally takes up **less space**
- can be **quicker** to produce, especially if you make a mistake or decide to edit it
- can easily be **stored** until required

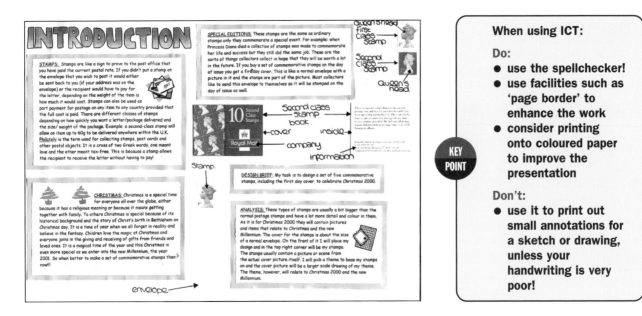

Fig 2.4 Example of illustrated written elements

Bubble charts and spider diagrams

An excellent starting point for an analysis of a problem is to **brainstorm**. Rather than just make a list, it can be more effective to present your thoughts in the form of a **diagram** that can allow you to:

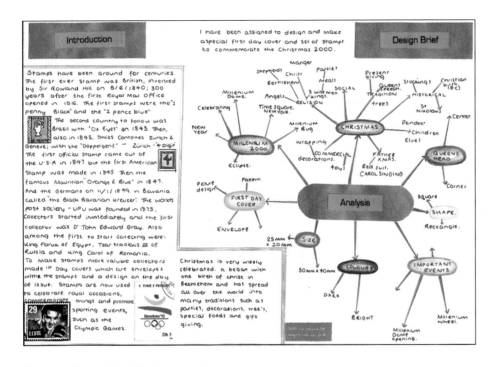

- **group items** together that are similar
- **create links** between items
- **highlight** important areas
- highlight omissions or possible areas for further consideration

Fig 2.5 Example of brainstorming

Freehand drawing (sketching)

> **You will find more information on graphic techniques in Chapter 4, Graphic products.**

The advantages of using **annotated sketches** have already been noted. Freehand sketching should be used at all times other than for final developments and working drawings, where it is important to be accurate.

Remember these points when sketching.

- Sketch in **three dimensions**. **Isometric** is often the most useful method, and **isometric grid paper** is readily available if you need help.
- Don't use soft pencils as these smudge easily.
- Start by lightly drawing the box the item would fit into (known as '**crating out**') and slowly put in the basic features – then darken the required lines and add rendering as necessary.
- Don't use **fine liners** unless you are confident and competent, except for **highlighting** important features.
- It is not always necessary to render fully, or even complete a sketch, as long as you have clearly communicated the point being made.

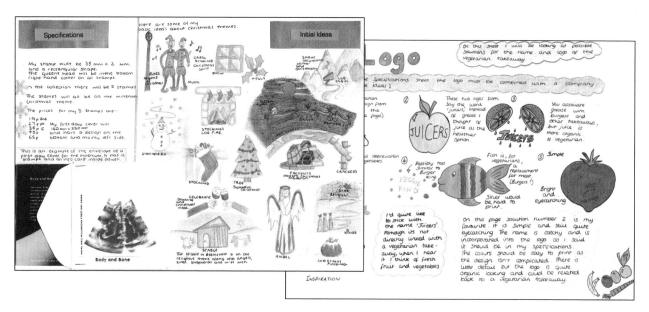

Fig 2.6 Annotated freehand sketching used to communicate ideas

Cut and paste

You will probably collect a lot of clippings, cuttings and pictures from magazines and leaflets during your coursework, which can usefully be used to **illustrate** your work to save even more time.

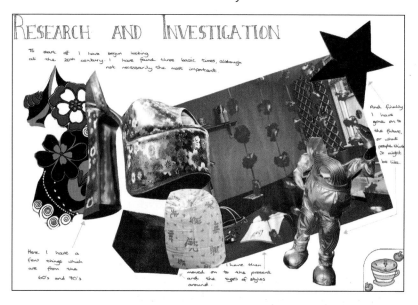

- **Cut and mount** the pieces carefully. A haphazard display may count against your communication mark.
- Be **selective**: only use directly relevant material.
- Always **annotate** your chosen illustrations or you will gain less credit.
- If you do not wish to destroy the original source material, e.g. a library book, then use a scanner or a photocopier. Remember to include the source in the **bibliography**.

Fig 2.7 Example of cut and paste using photographs and clippings

Photographs

Although expensive, time-consuming and sometimes frustrating in terms of quality, **photographs** allow you to present exactly what you want and are good evidence of **primary research**.

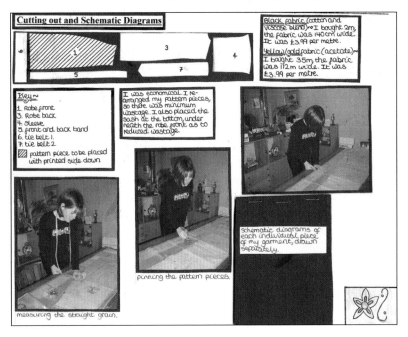

Fig 2.8 Photographs used to record coursework elements

They are extremely useful as a method of creating a permanent record of work that may otherwise be lost:

- **Perishable** items such as food
- **Stages** in construction
- **Changes** made during development
- Testing to **destruction**

Digital cameras are especially useful, if available, as the images are easily imported into computer programs.

When presenting the questionnaires, only include the original or a representative example in the main body of the folder. The rest should be in the appendix, annotated, if necessary, to highlight the relevant responses.

Questionnaires

These should be used to gather a wide range of **opinion or facts** about a particular aspect of your project. It is a particularly useful method to use during the **research and evaluation** sections of your folder.

Follow these guidelines when using a questionnaire.

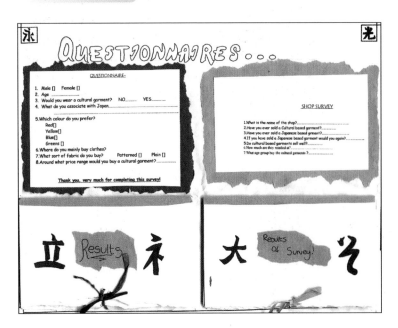

Fig 2.9 Example of questionnaires used in coursework

- Seek the views of people who have some **direct connection** with the topic concerned or your results may be misleading.
- Be **clear** what it is you intend to find out.
- Decide **how many** people you need to ask.
- Start with **'yes/no'** questions.
- **Multiple choice** questions must have a space for 'other' answers.
- Plan the questionnaire so that it is in a **logical order** and does not take too long.
- Allow space(s) for **comments**.
- **Do not** ask for **personal** information.
- **Thank** the people who fill it in!

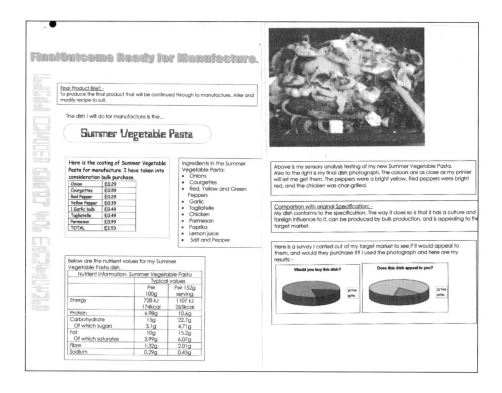

Fig 2.10 Charts and graphs used to display data analysis

Charts, tables and graphs

You will generate a lot of information from surveys, questionnaires and tests. The best way to present this information is through charts, tables and graphs. These can be easily produced **graphically**, where personalising the presentation can enhance them.

KEY POINT

If you have a **large quantity of data**, it would be more appropriate to use **ICT** to produce your spreadsheets, tables and charts/graphs. This will provide further evidence of your skill in the use of ICT.

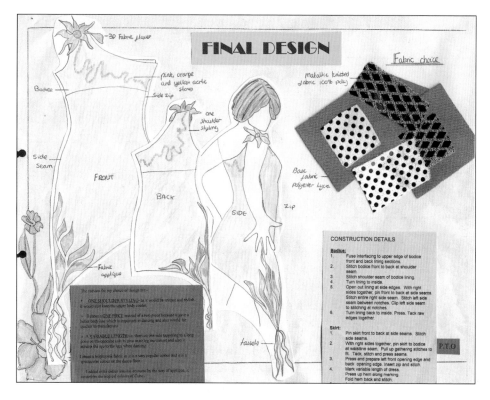

Fig 2.11 Example of working drawings

Working drawings

The main purpose of these is to give all the information required to realise your design. Any method of drawing is acceptable, but the following **must** be included.

- Dimensions
- Materials
- Construction details
- Parts list
- Finishes
- Quality control, e.g. tolerances

Gaining more space

It will often make sense to display all the work relating to a single topic on one sheet of paper. This can be achieved by the use of **pockets, folding, layering and sectioning**.

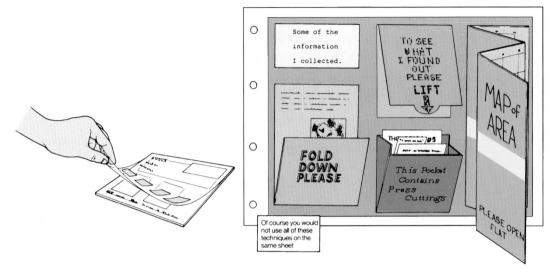

Fig 2.12 Methods of gaining space

Mistakes

You will always make mistakes during your project. Correct small errors as **neatly** as possible, but avoid large areas of crossing out or the use of correcting fluid, as it looks unsightly.

> **KEY POINT**
>
> Never discard a sheet that has a mistake on it. Cut and paste a corrected version by:
> - making a feature of it by double mounting
> - cutting and pasting the correct portion onto another sheet for finishing off

Choosing a situation or need

> **KEY POINT**
>
> This is the starting point for all coursework and is important because it forms the introduction to your project. In it you must demonstrate to the examiner that you are to be engaged in a real problem which is worthwhile and not contrived.

The design problem might be identified from any of the following ideas.

- An activity, interest or hobby you are involved with
- A need generated by the community, local business or school
- Something set by your teacher or suggested by the examination board

The introduction could include the following information.

- Some background information on your involvement
- Details of how the need has arisen
- A recent history of the problem
- Why you consider it a worthwhile activity

This section should be kept short, serving only to set the design brief in context. It generates few marks but must demonstrate your understanding of the need.

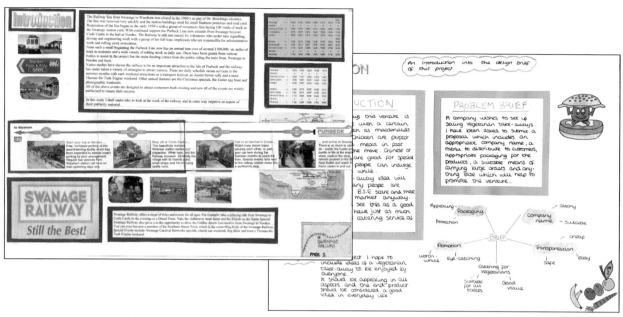

Fig 2.13 Introducing the project

Preparing a design brief and specification

KEY POINT

The design brief should:
● be a short statement of the problem to be solved
It should not:
● describe the solution or outcome

Present the product specification as a numbered list, or a series of bullet points, in order to make it easier to evaluate the solution.

The following should be considered in your product specification.

● **Context**
  ● Where and how will the product be used?
  ● Who will use it?
  ● What are the implications for the selection of materials?

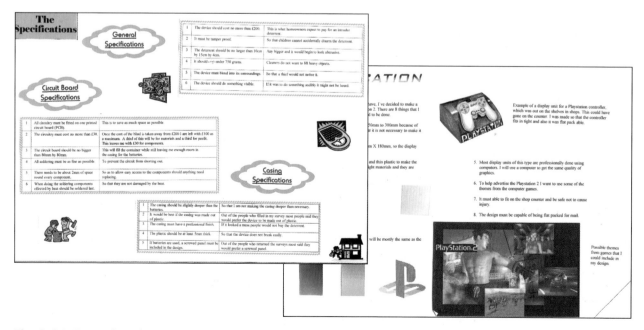

Fig 2.14 Examples of project briefs and specifications

- **Aesthetics**
  - Colour and texture
  - Styling, feel, taste, smell
  - Size, shape, proportion, balance
- **Performance**
  - Function – how it works and why
  - Ergonomic/anthropometric considerations
  - Secondary functions – quality, reliability, safety, fashion, efficiency

- **Industrial practices**
  - Costing
  - Scale of production
  - Quality control/assurance
  - Timescale
- **Maintenance**
  - How will it be maintained?

Analysis and research

Analysis is a '**question asking**' activity, and can be carried out at any stage of the process as necessary. It is a way of establishing what information you still need to find out in order to proceed at any stage of the project.

Research is the means by which you find out this information and is important because it provides the **knowledge and understanding** needed to make **decisions**.

Follow these guidelines.

- Include only **relevant** information.
- Use a wide **variety** of sources.
- Write **letters**, make **phonecalls**, send **e-mails** to people who may have relevant knowledge (see Fig 2.16).
- **Critically** analyse the information you acquire.
- Always draw **conclusions** from your research.
- Go back to your initial specification and **adapt** as necessary.

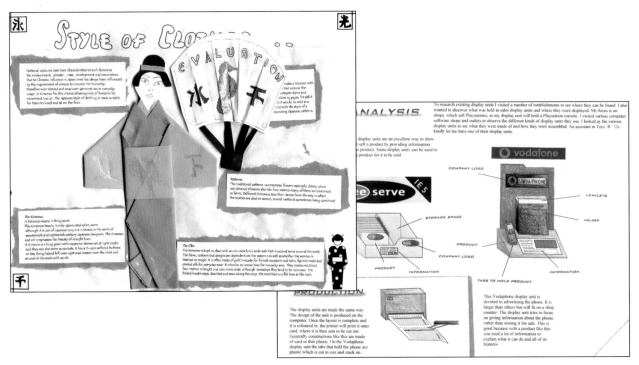

Fig 2.15 Examples of analysis and research

School Name
Education Road
Town
County
Postcode

Date

Named Person
Relevant Company
Industry Road
Town
County
Postcode

Dear Sir/Madam

I am a Year 11 student studying GCSE Technology and at present I am working on my coursework project which is concerned with . . .

I am writing to you because of your company's involvement in this field and I hope that you will be able to answer the following questions that may help me with this project.

I am particularly interested in the following aspects:

1. Question
2. Question
3. Question

I would like to thank you in advance for your consideration and for any information that you can give me. I enclose a stamped addressed envelope.

Yours faithfully

GKO Elphinstone

G.K.O. Elphinstone

I confirm that G.K.O. Elphinstone is currently engaged on this project for his GCSE Technology coursework, and any assistance you could give him would be very much appreciated.

Signed .... *R. Davis* .... Technology teacher

Fig 2.16 Writing letters

If the school policy allows, use the school's own **headed paper**. Otherwise, use the school's address with permission.

Write as **early** as possible – replies can be slow!

It is an advantage to write to a particular person. If you do not have a name, write to a post-holder connected with your project. If in doubt, write to **'Customer Relations'**.

Say who you are and what you are doing.

Say why you are writing to them in particular.

Do not ask them to solve your problem!

Ask a **few specific** questions to which you believe they should reasonably know the answers.

Do not ask commercially sensitive questions.

Thank them in advance and write again if they reply. **Send a large SAE.**

Sign the letter clearly.

Ask your **teacher to countersign** the letter as this can increase your chances of a reply.

Preparing a product analysis or case study

It is often worthwhile to study products similar to your own in detail, not only in order to focus your research more clearly, but also as a means of covering **industrial practices**. A typical case study would contain some or all of the following elements.

- **Product details**
  - The name of the product, including a picture if appropriate
  - Information regarding cost, availability, advertising
  - An explanation of its function, including safety and hygiene
- **Visual analysis**
  - An annotated sketch of the product
  - An exploded pictorial
  - Detailed drawings of significant parts
- **Materials analysis**
  - Details of the raw materials and parts used
  - Possible reasons for their choice
  - Details of finishes used
- **Production analysis**
  - Details about how the product and its parts are manufactured
  - Possible reasons why these methods are chosen
- **Marketing analysis**
  - Details of how the product is marketed
  - Discussion about the intended market
  - Comparison with similar products on the market

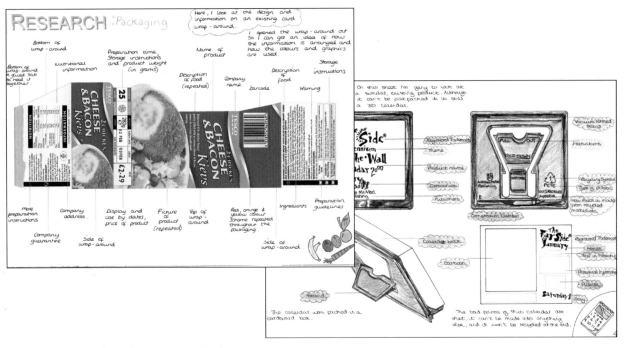

Fig 2.17 Examples of product analysis

> A good model for this would be the *Which? Reports* published by the Consumers' Association.

- **Evaluation**
  - A critical analysis of the product, commenting on issues such as aesthetics, ease of use, safety, durability, environmental considerations, value for money.

> **KEY POINT**
>
> **Knowledge of industrial practices** is a major assessment objective of the coursework element. If you do not cover it by looking at existing products in the research section, you must do so once you have designed your own product.
>
> It would be advisable, if time allows, to check at least the materials and production analysis sections for your own product.

Generating initial ideas

In order to produce an effective solution, you must consider a wide range of alternatives. Each of these must be **compared with the specification** to identify the most **promising** for further **development**.

● Produce your ideas in **sketch form** – the ideas are more important at this stage than technical accuracy.

● **Compare and contrast** your ideas by reference to your specification in note form.

● **Explain** why you intend to develop some ideas and reject others.

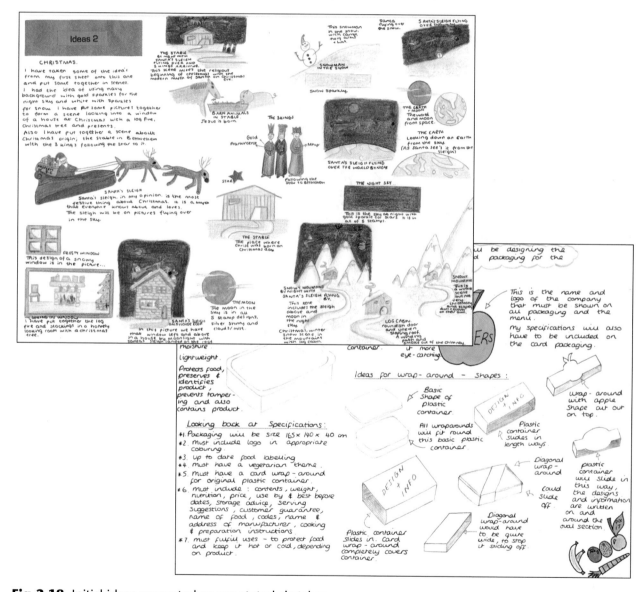

Fig 2.18 Initial ideas presented as annotated sketches

Developing the solution: modelling and prototypes

This involves a clearly **reasoned** development of an accepted idea, or group of ideas, into a final **solution**. You must **modify** your proposal progressively, **recording** the changes and the **reasons** for them, to ensure a sound solution when viewed against the specification.

See Section 4.3 in Chapter 4, Graphic products for information about modelling.

- Test ideas as **mock-ups** as you go along using modelling materials.
- Seek the opinions of **clients/experts/potential users**.
- Draw upon earlier research to aid decisions on materials and manufacturing methods.
- Consider the social, environmental and health and safety aspects of your design.
- Sketching should now give way to more **detailed** forms of drawing.
- The solution should be presented as a **working drawing**, including parts lists and manufacturing details. It may be advantageous to use ICT to produce the working drawings.

KEY POINT

This assessment objective carries a significant number of marks and therefore should not be simply a working drawing of your favourite idea. The process of developing the idea into a working product or prototype is what you should concentrate on presenting in detail.

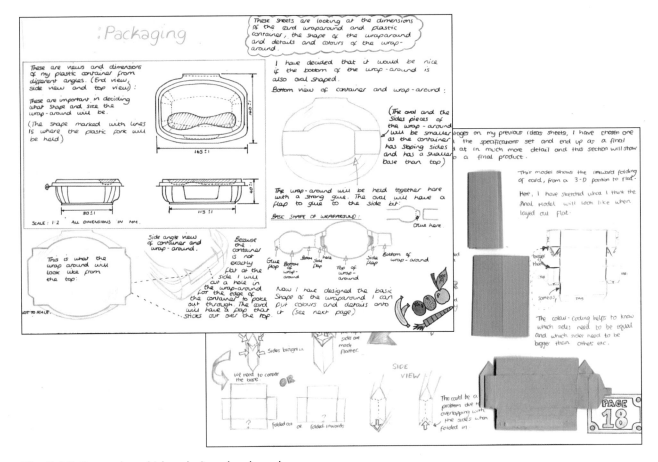

Fig 2.19 Examples of ideas being developed

Planning for production

KEY POINT

You should not start to make your product without first planning it out carefully. This involves consideration of all the necessary tools, materials and processes required to produce a high quality product safely. Reference must be made to the form in which the materials are produced for manufacture in industry, and you must take account of the implications for your own design.

Help on presenting flow charts and sequence diagrams can be found in Chapter 4, Graphic products.

When producing a planning schedule, you must identify:

- each distinct **stage** in the manufacturing process in consecutive order
- the **time** available for each stage
- the tools, techniques, equipment and machinery needed
- ways of ensuring **quality control**
- health and safety issues

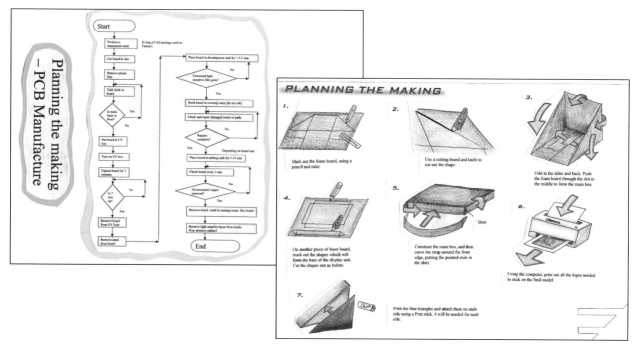

Fig 2.20 Planning schedules used in projects

> **KEY POINT**
>
> **Although a successful outcome is evidence of good planning, you should record any deviations and modifications made which became necessary during the process, noting the reasons for the changes to the original plan.**

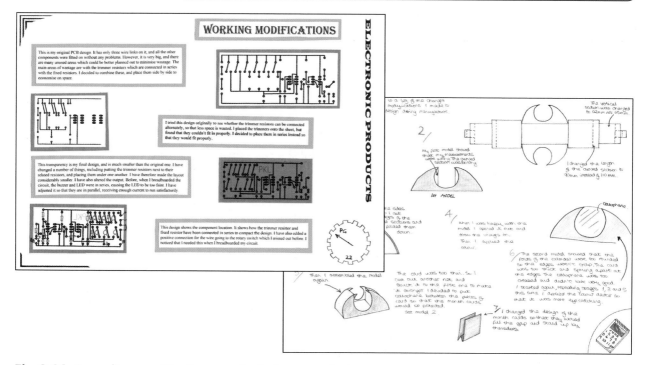

Fig 2.21 Recording modifications made during manufacture

Manufacturing the solution

This is the culmination of your project and as such must be given the time needed to ensure a quality outcome.

Think twice – act once!

- **Practise** new skills first.
- Carry out **test pieces** to ensure suitability of material and process.
- Make adequate provision to store your work safely.
- Seek **advice** at every stage.

 KEY POINT

It is acceptable to get specialist help with processes that your school cannot cater for. However, you must state the extent of such help on each occasion.

As this help cannot be credited in terms of marks, you must ensure that you have enough evidence of your own making skills.

Fig 2.22 Examples of completed coursework products

Testing and evaluation

Include comments on further developments that have come to light during the process of manufacture and testing.

You will have evaluated your project at various stages as an ongoing exercise – known as **formative evaluation**. Each time, you should have given reasons for how you have proceeded. You must now analyse your outcome critically in relation to your brief and specification, along with any issues arising from the making – known as **summative evaluation**.

Do:

- Produce an evaluation point for each of the specification bullet points.
- Test the solution in its **intended setting** – taking photographs if possible.
- Seek the **opinions of others** – especially those specialists you sought advice from earlier in the project.
- Suggest **modifications and improvements**.

Don't:

- Merely write an account of how you got on, like an extract from a diary.
- Ask opinions of classmates, friends and relations unless they have significant/relevant knowledge that can assist your evaluation.

KEY POINT

Be critical – with reasoning.

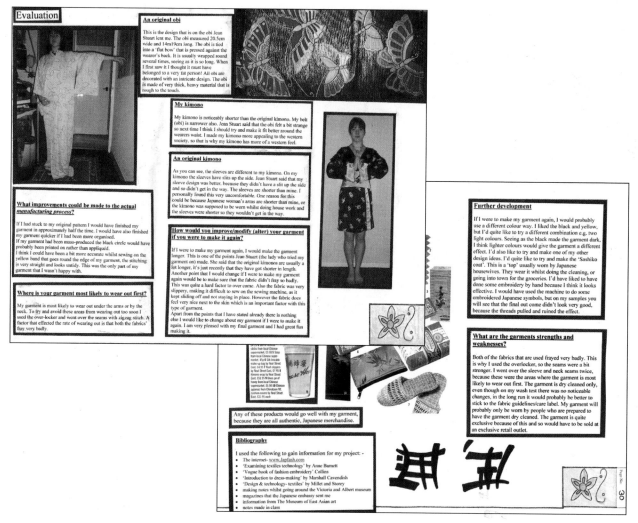

Fig 2.23 Examples of evaluations

3 Food technology

The following topics are covered in this chapter:

- Equipment
- Nutrition
- Food preparation
- Product development
- Packaging

3.1 Equipment

LEARNING SUMMARY

After studying this section you should be able to:

- *select, name and safely use pieces of equipment for a particular task or process*

Applications of equipment

AQA
EDEXCEL
OCR
WJEC
NICCEA

Whilst it is likely that you will only have had direct experience of working in a domestic/school kitchen – the equivalent of a 'test kitchen' – you should also understand the applications on an **industrial** scale. This will involve the **scaling up** of your product for **manufacture** in terms of **quantities** and using **industrial sized equipment**.

Equipment	Comments	Appearance
Teaspoon, dessertspoon and tablespoon	• Used for measuring, mixing, tasting and eating	
Kitchen knives	• Used for cutting and scraping • Stainless steel knives are best and should be kept sharp for safety reasons	
Kitchen scissors	• Used for cutting greaseproof paper and ingredients such as parsley	
Peelers and corers	• Used for removing peel and cores from fruit and vegetables • More effective than knives, especially for the less experienced	Apple corer Potato peeler

Fig 3.1 Food preparation tools and equipment

Equipment	Comments	Appearance
Graters and choppers	• Useful for smaller amounts of food and in the absence of a processor • Require careful cleaning	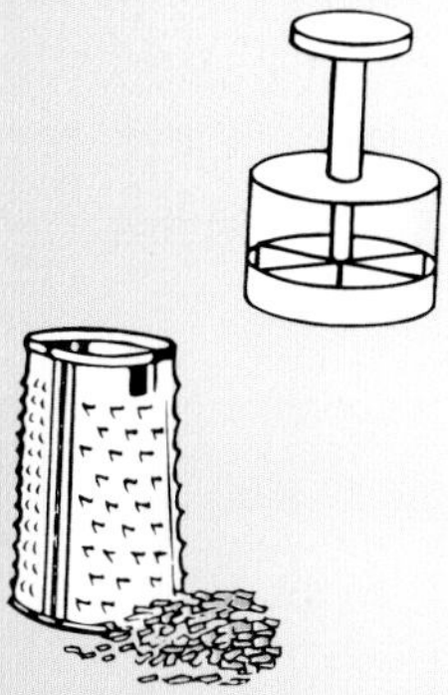
Wooden spoon	• Used for stirring, beating and blending – especially when other utensils are not scratch resistant • Heat resistant • Available in various shapes and sizes	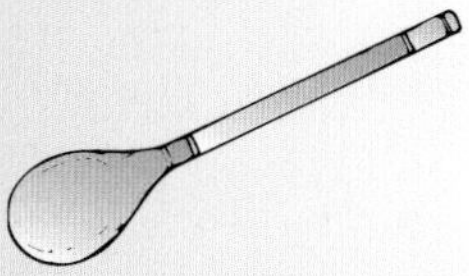
Hand whisk	• Used on small quantities of ingredients to add bulk • More effective than a spoon	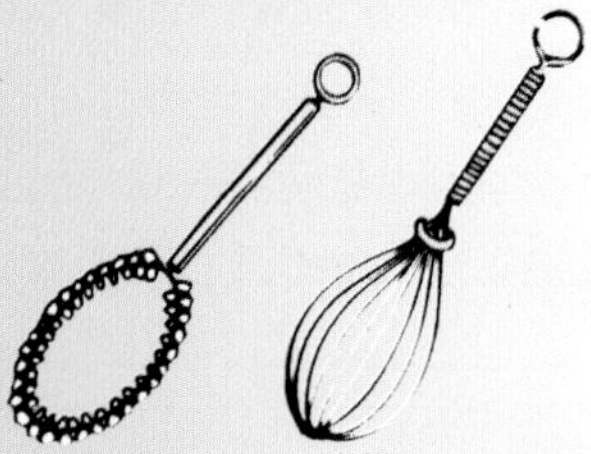
Rotary whisk	• More convenient for aerating larger quantities – especially when an electric mixer is not available	
Electric food mixer	• Essential for larger quantities of ingredients • Much quicker and more effective than the hand versions • Can be used for mixing as well as aerating	
Food processor	• Used for grating, slicing, mincing, mixing, and kneading dough • The various fittings require thorough cleaning, with special care needed for the blade attachment	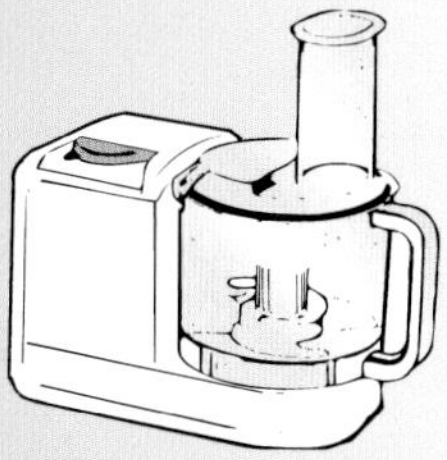
Scales	• Used for measuring ingredients • Greater accuracy is required for smaller quantities, with electronic scales now becoming more popular	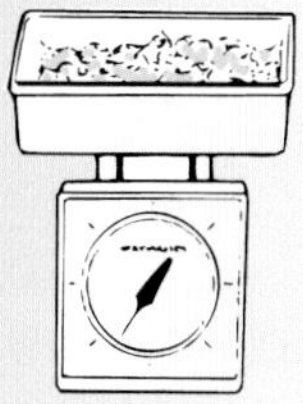

Fig 3.1 Food preparation tools and equipment (continued)

Equipment	Comments Appearance	
Measuring jug	• Used for measuring the volume of liquids	
Measuring spoons	• Used mainly for the measuring of dry ingredients • Available in various commonly used sizes	
Spatula and palette knife	• Used for spreading or cleaning out bowls	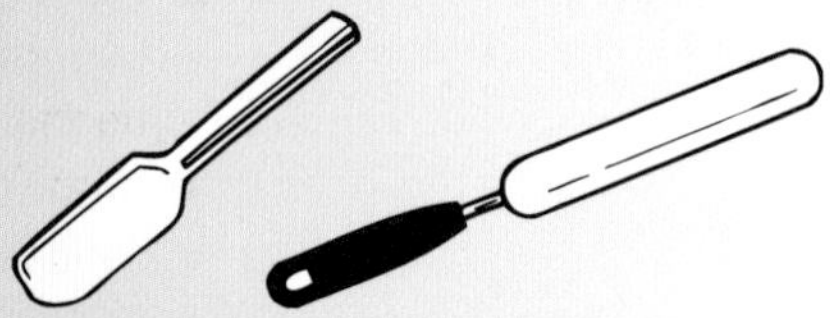
Mixing bowls and basins	• Used for mixing dry and wet ingredients • Made from earthenware, glass or plastic and available in a wide range of sizes	
Chopping board	• Used for cutting up ingredients and to protect the work surfaces • Traditionally made from wood, but much more hygienic if made from a synthetic material	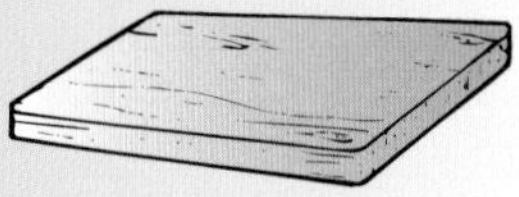
Rolling pin and cutters	• The basic equipment for pastry and biscuit making, used in conjunction with a chopping board • Available in various materials and more effective if kept cool	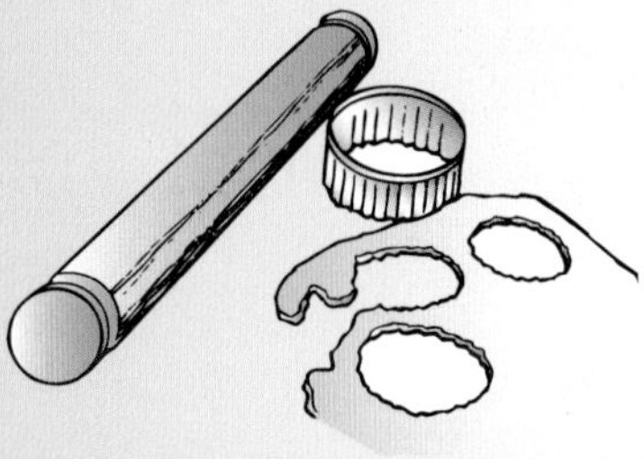
Saucepans	• Available in a wide range of sizes and materials • Lids should fit well when provided and ovenproof handles should be chosen for use in an oven	
Frying pan	• Used for shallow frying with small amounts of fat or oil, and can be more effective when used with a lid • Those with more rounded sides are easier to use when removing omelettes, etc	

Fig 3.1 Food preparation tools and equipment (continued)

Equipment	Comments	Appearance
Casseroles and dishes	• Available in various sizes and made from ceramics, glass or metal • Metal must not be used in a microwave oven	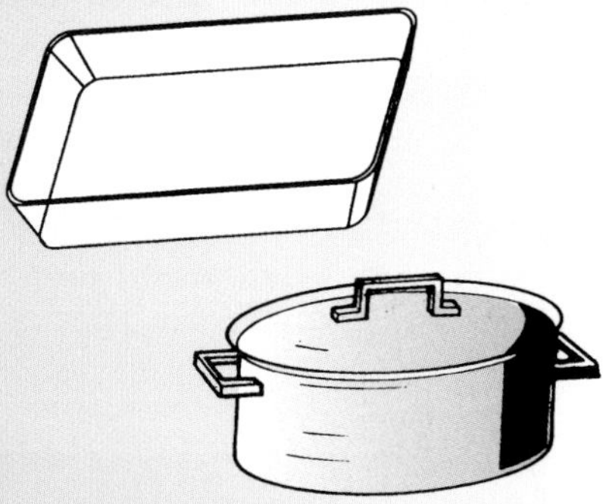
Microwave oven	• Useful for rapid cooking of frozen and pre-prepared or convenience foods • Should be serviced regularly for safety reasons	
Deep fat fryer	• Much safer than a large saucepan on a hob	
Pressure cooker	• Used for preparing stews and vegetables • Saves time and fuel	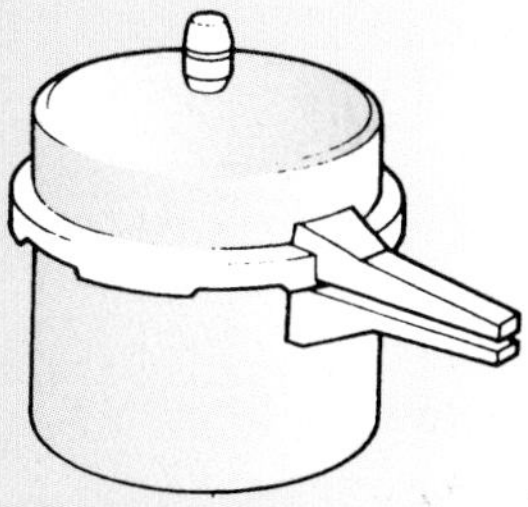
Conventional oven	• Can be fuelled by electricity or gas, with many now being fan-assisted to increase efficiency • An obvious safety hazard because of the high temperatures, and must also be regularly serviced	

Fig 3.1 Food preparation tools and equipment (continued)

KEY POINT Kitchen equipment should also be looked at for its ergonomic design as well as its aesthetic appeal in terms of colour, shape and feel.

3.2 Nutrition

After studying this section you should be able to:
- *understand the importance of nutrients*
- *understand dietary needs*

Nutrients

AQA
EDEXCEL
OCR
WJEC
NICCEA

Food contains the **fuel** required by our bodies to produce the **energy** needed to sustain them at work, rest and play. Food also contains elements that are not needed and can be actually **detrimental** to health.

The components that are useful to the body are known as **nutrients.** These are classified into **five** different types and have different jobs within the body.

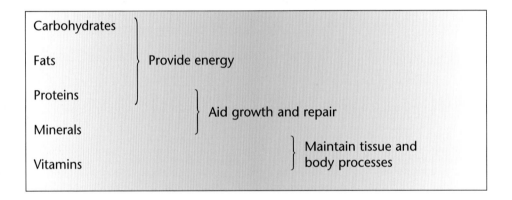

Carbohydrates

Fats } Provide energy

Proteins

Minerals } Aid growth and repair

Vitamins } Maintain tissue and body processes

Carbohydrates

Carbohydrates are produced by plants through **photosynthesis** and contain the three elements: **carbon, hydrogen and oxygen**.
The simplest forms are **monosaccharides**, which are sugars that are **absorbed** into the body.

 KEY POINT   **Examples of monosaccharides are** fructose, glucose and galactose.

When photosynthesis continues, **disaccharides** are formed from two monosaccharides. These are broken down by **digestion** into monosaccharides before being absorbed into the body.

 KEY POINT   **Examples of disaccharides are** sucrose, maltose and lactose.

Further photosynthesis produces the third main group of carbohydrates, **polysaccharides**. These again have to be broken down into **glucose** during digestion before being absorbed.

There are five main polysaccharides.

- **Starch** forms the basis of most diets around the world. Each area has its own **staple diet**, e.g. rice in Asia, potatoes in Europe and corn in South America.
- **Dextrin** is formed when starch foods are baked/toasted.
- **Cellulose** forms roughage in our diet, being derived from plant cells.
- **Pectin** gels in water and therefore aids setting.
- **Glycogen** is stored in the liver, having been formed during digestion. It is then converted to glucose.

The most common of these are the starches which, unlike sugars, are **not soluble** in water. They are also known as **complex carbohydrates**, being made from more complex molecules than sugars.

Fats

Fats (including **oils**, which generally remain liquid at room temperature) are needed by the body in far **smaller quantities** than carbohydrates and are obtained from **milk, oily fish, meat, seeds and nuts**. They are made up of the same three elements as carbohydrates, and perform the similar functions of providing **energy, insulation** when stored, and **transporting vitamins**.

> **KEY POINT**
> The name for a fat is a triglyceride, **which is formed from a molecule of glycerol and three chains of** fatty acids.

There are three fatty acids.

- **Saturated fatty acids** – all the carbon atoms in the chain have a **single link** or **bond**
- **Mono-unsaturated fatty acids** – there is a **single double bond** in the chain, enabling it to take on more hydrogen atoms, **not saturated**
- **Poly-unsaturated fatty acids** – there is **more than one double bond** and a greater capacity to take on hydrogen atoms

Most fats and oils contain saturated and unsaturated fatty acids. Those containing a **high proportion** of **unsaturated** fatty acids are softer at room temperature because they have a **lower melting point**.

- Animal fats contain a higher percentage of saturated fatty acids.
- Vegetable oils are high in unsaturated fatty acids.

This is critical for health reasons since **saturated fats** can **increase the cholesterol level** in the blood, which can lead to **heart disease** if cholesterol is **deposited** on the walls of the arteries.

It is a mistake to suggest that fats should be removed altogether from a diet, as they are essential to the healthy functioning of the body.

Proteins

Proteins are made up of the four elements: **carbon, hydrogen, oxygen and nitrogen**. In addition, some proteins contain **sulphur** and **phosphorus**.

> **KEY POINT**
> These elements form chains of protein molecules called amino acids. They **are essential to life and can only be obtained from food.**

Adults require **eight** essential amino acids, whilst **children** require a **further two**.

- **High biological value proteins (HBV)** are foods that contain **all ten** essential amino acids. They are generally found in **meat, fish, eggs and milk**.
- **Low biological value proteins (LBV)** are foods that have **one or more** essential amino acids **absent**. These are plant food such as **pulses, nuts and cereals**. However, they can supplement each other in combination.

You should emphasise that the best way to ensure that the body has the correct amounts of essential amino acids is by having a variety of foods.

> **KEY POINT**
>
> **Soya beans, however, are HBV and form the basis of** textured vegetable protein (TVP), **a common meat substitute used by** vegetarians.

Minerals

Minerals play an important part in **forming and maintaining**:

- the skeletal structure of the body, e.g.
  - **calcium** in milk, cheese and tofu
  - **phosphorus** in cheese, eggs, fish, meat, cereals, nuts and seeds
  - **magnesium**
- the soft tissues, e.g.
  - **potassium** in fruit, vegetables and whole grains
- the body fluid, e.g.
  - **sodium** in salt
  - **chlorine**

Iron, found in liver, kidney and egg yolk, is needed for the replacement of red blood cells. Because of menstruation, women require more iron than men do.

Iodine, a trace element found in seafood, milk and spinach, is needed in small amounts to help regulate the metabolism of the body.

Fluorine, found in seafood and the water supply in some areas, is another trace element required in small amounts to harden the enamel in teeth.

Vitamins

Vitamins are found in small quantities in many foods and can be put into **two groups**.

- Fat-soluble – A, D, E and K
- Water-soluble – B group and C

Vitamin A (Retinol or carotene)	• Good for the eyesight, healthy skin and mucous membranes • Found in dairy produce, eggs, oily fish, leafy vegetables, and especially carrots
Vitamin D (Calciferol)	• Good for the formation of bones and teeth • Found mainly in animal foods – especially fish liver oils
Vitamin E	• Helps the functioning of the lungs and red cell membranes • Found in eggs, cereal oils, liver and meat
Vitamin K	• Stops bleeding by clotting blood • Found in green vegetables, fish and liver

Vitamin B1 (Thiamin)	• Helps release the energy from carbohydrates and is needed for the functioning of the nervous system • Found in many foods, especially wholegrain cereals
Vitamin B2 (Riboflavin)	• Helps release energy and therefore aids growth rate in children • Found in many animal products, dairy produce, green vegetables, nuts, seeds and yeast
Vitamin B3 (Niacin)	• Helps release energy • Found in many foods
Vitamin B12	• Aids the metabolism of amino acids • Found only in animal products
Vitamin C (Ascorbic acid)	• Makes connective tissue, aids the absorption of iron and assists in the formation of bones and teeth. Protects against infection and allergies • Found mainly in fruit and raw green vegetables

It is important to take care when preparing food containing the water-soluble vitamins as air and heat easily destroy them.

Fibre (roughage)

Dietary fibre is not digested and absorbed by the body, being mainly the **outer husks and peel** of plant food. However, it is important to the **digestive system** because it can absorb water and adds bulk, thus helping to **avoid constipation**. It is also known as **NSP** (non-starch polysaccharide).

Water

Water is **essential for life** and is contained in fruit, juices, vegetables and milk. There is a constant requirement for water as it is lost through perspiration, breathing out and urine, which removes some of the body's waste products in solution.

Dietary needs

AQA
EDEXCEL
OCR
WJEC
NICCEA

Metabolism

The rate at which the body uses up energy is known as the **metabolic rate (MR)**. This rate varies depending on the height and weight of the person, their age and what they are doing. The rate at which the body uses energy to perform the basic tasks of remaining alive is known as the **basal metabolic rate (BMR)**.

Dietary reference values (DRVs)

Because of the varying requirements of different groups of people in the community, the Department of Health estimated their needs and published standards for the amounts of nutrients required known as Recommended Daily Intakes **(RDI)** or Amounts **(RDA)**.

Further research generated a report in 1991 by the Committee on Medical Aspects of Food Policy **(COMA)**, which recommended amounts of energy, protein, fat, carbohydrate, vitamins and minerals for different groups of people. The Dietary Reference Values are given in the next table.

Lower Reference Nutrient Intake (LRNI)	The minimum requirement for a nutrient needed by those with the lowest needs – usually about 3% of the group concerned
Estimated Average Requirement (EAR)	The average daily need for a particular nutrient for a specific group of individuals
Reference Nutrient Intake (RNI)	The requirement for a nutrient that will satisfy the needs of 97% of the target group
Safe Intake	Where there is not enough information concerning nutrients and certain groups, this value is given as being enough for most people without producing undesirable side effects.

> **You should concentrate on the concept of a balanced diet providing the right amount of nutrients.**

KEY POINT

Use ICT to help. There are a number of computer programs that will analyse a recipe based on the weights of constituents, providing a printout of DRVs for specific target groups. This allows you to adjust the amounts before proceeding with further product development.

Food selection	Amount (grams)
mushrooms: raw	50
omelette	230
onions: raw	150
milk: semi-skim pasteurised	125
cheese: Cheddar low-fat	50
tomatoes: raw	120

Dietary Reference Values

% Energy – protein	27
% Energy – Fat	61
% Energy – carbohydrate	12

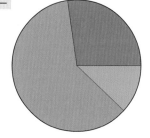

> **You will only gain marks for using nutrient analysis programs in coursework if some explanation and evidence of application is also given.**

DRV based on requirements for a boy aged 15–18 years

DRV ➡ ⬅ 100%

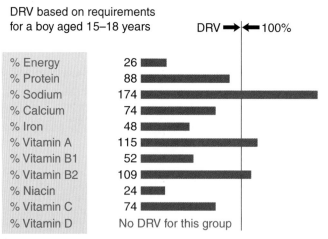

% Energy	26
% Protein	88
% Sodium	174
% Calcium	74
% Iron	48
% Vitamin A	115
% Vitamin B1	52
% Vitamin B2	109
% Niacin	24
% Vitamin C	74
% Vitamin D	No DRV for this group

Fig 3.2 Example of computer analysis of DRVs

Special dietary needs

Diet changes throughout the world through custom and availability. However, diet is also changed through **necessity**, e.g. medical reasons, or by **choice**, e.g. religion, personal beliefs.

The following are some of the dietary preferences/needs that exist.

Diabetics	<ul><li>Must control carbohydrate intake by eating low-fat, high-fibre foods</li></ul>
Pregnant women	<ul><li>Need high calcium and iron intake, but not from liver</li><li>Need protein-rich food</li><li>Should consume more fresh fruit and vegetables</li><li>Should avoid alcohol</li></ul>
Vegetarians • Lacto-ovo-vegetarians • Lacto-vegetarians • Vegans	<ul><li>Do not eat food that involves killing animals and the by-products of slaughter, e.g. rennet, gelatine</li><li>Eat both dairy products and eggs</li><li>Eat dairy products but not eggs</li><li>Refuse all animal products</li></ul>
Muslims	<ul><li>Do not eat pork and have varied views on fish</li><li>Only eat meat slaughtered and prepared in a ritual way – Halal</li><li>Do not consume alcohol</li><li>Fast on a regular basis</li></ul>
Hindus	<ul><li>Do not eat beef</li><li>Are generally vegetarians, eating mostly vegetables</li><li>Do not consume alcohol</li></ul>
Sikhs	<ul><li>Similar to Hindus, with some eating mutton or chicken</li></ul>
Jews	<ul><li>Do not eat pork</li><li>Do not eat shellfish – only fish with scales and fins</li><li>Meat must be ritually slaughtered – kosher</li><li>Do not eat milk and meat together</li></ul>
Rastafarians	<ul><li>Do not eat animal products except milk in some cases</li><li>Avoid canned or processed foods</li><li>Do not consume salt, coffee or alcohol</li><li>Eat only organically grown food</li></ul>

Meat substitutes

Vegetarians and people with health problems often require meat substitutes. **Soya** products have long been used in this respect, e.g. **textured vegetable protein (TVP) and tofu**.

More recently Quorn, a **mycoprotein**, has been developed. This fungus is low in saturated fat and is a good source of protein and fibre.

3.3 Food preparation

After studying this section you should be able to:

- **understand the importance of hygiene**
- **understand the causes of food poisoning**
- **understand the working properties of food**
- **understand food processing**
- **understand the methods of food preservation**

The aim of food preparation is to produce food that is:

- attractive and nutritious to eat
- safe to eat

Food hygiene

AQA
EDEXCEL
OCR
WJEC
NICCEA

For food to be safe to eat it is important that it is prepared, cooked and stored hygienically.

Good practice depends on a number of factors, all of which must be observed to the highest standards possible.

Raw materials

See Section 3.5, Packaging for further information about labelling.

- Do not use materials that are in any way **unfit for consumption**.
- When buying food, do so from establishments that are **well run** and that provide **fresh products** that have been stored properly.
- Carefully check any **date marks** that should be displayed.
- Observe any **product instructions** regarding storage and preparation.

Water quality

As water is used throughout the preparation and cooking process, it is essential that an adequate **uncontaminated** water supply is available.

Personal hygiene of food handlers

The personal cleanliness of all food handlers must be maintained to the highest standards, including:

- wearing **clean overalls**, tying back hair and using head coverings
- **washing hands** in hot, soapy water before and during handling food – especially between handling cooked and uncooked foods or after using the toilet
- **not smoking** when preparing food or in the vicinity
- not working when **ill** and protecting cuts and grazes adequately

Kitchen hygiene

When working in a kitchen you should take steps to:

- maintain **clean equipment and surfaces** at all times by regular washing with hot, soapy water
- **change** kitchen cloths regularly and **bleach** them when they are washed
- use **anti-bacterial sprays** where possible
- **empty** rubbish bins regularly to avoid attracting pests

Although they are considered part of the family, pets must also be kept away from food preparation areas.

In addition, business premises should:

- be clean and in **good repair**
- have an adequate **water supply** – cold and hot
- have adequate **sanitary facilities**, lighting and ventilation
- have suitable **pest control** arrangements
- be **designed** to take account of good **hygiene practices**

Preventing food contamination

Food must be stored properly and **cross-contamination** avoided between cooked and raw foods – either by **contact** or the **utensils used**.

> **KEY POINT**
> Commercially, the maintenance of food hygiene requires a knowledge and application of the regulations and laws, staff training and quality control systems.

Food poisoning

AQA
EDEXCEL
OCR
WJEC
NICCEA

The outcome of poor hygiene is that food can become infected at any stage of its production, leading to food poisoning.

> **KEY POINT**
>
> **Food is considered to be contaminated if it is found to contain infective viruses or bacteria or foreign bodies such as chemicals, metals or poisonous plants.**

Bacteria

There are many types of bacteria, some of which will cause food spoilage and are known as **food spoilage bacteria**. These are unlikely to cause food poisoning. The few bacteria that do are known as **pathogenic bacteria**. Bacteria require a number of conditions to be present to grow.

- **Temperature** – the pathogenic bacteria are most active around 37°C, the body's natural temperature. However, the **danger zone is 3–67°C**. Below this temperature most will become inactive (dormant), whilst above this temperature most will be destroyed.
- **Food** – the pathogenic bacteria do well in foods that are moist and high in protein.
  - High risk foods include cooked meat and poultry, seafood, dairy products, cooked rice, raw egg products, soups, stews, etc.
  - Low risk foods are those containing high amounts of salt or sugar, or which are acidic.
- **Moisture** – the bacteria require moisture to grow, but the absence of moisture in the food does not mean the absence of bacteria.
- **Oxygen** – most bacteria are **aerobic**, as they require oxygen to grow. However, some are **anaerobic** and can multiply without oxygen.
- **Time** – the bacteria need the right conditions to be available for at least the time it takes for them to divide – the **generation time**. Commonly this is about twenty minutes, which means that it will only take a few hours to produce enough bacteria to cause food poisoning.

See Food preservation, p.49 for more detail.

Bacterial food poisoning can be put into three main classes.

- **Infectious** – where the bacteria are parasites that cause illness.
- **Toxic** – where the bacteria's waste products (**toxins**) poison the body.
- A **combination** of these two.

Yeasts

Yeasts use the sugar in foods to reproduce and give off carbon dioxide (a process known as **fermentation**). Although this is useful in such processes as bread and wine making, it can spoil other foods – especially those with high sugar contents.

Moulds

Moulds grow when the spores of fungi fall on the surface of foods. Some, as in blue cheeses, are encouraged, but others will produce harmful **toxins**.

Enzymes

Enzymes, although not **micro-organisms** like bacteria, yeasts and moulds, are chemicals that change food such as fruit by partially digesting it, making it go brown and destroying the vitamin content. **Food spoilage bacteria** can produce such enzymes.

Working properties

Aerating

Aeration is the addition of gas to make the food **lighter in texture**. This can be achieved in a number of ways.

Whisking and beating	Adds air to such ingredients as eggs
Sieving	The action traps air between the particles
Kneading, folding and sealing	Again, traps air – especially in pastry products
Adding raising agents	Such as yeast, which adds carbon dioxide
Adding liquids	Such as milk, which adds steam during cooking

Binding

Some ingredients will not bind together when mixed and require another ingredient to aid this. Ingredients commonly used include **water, eggs, fats and flour**.

Bulking

In some recipes one ingredient will be used as the main part to **fill it out**. Examples are:

- **Flour** in bread
- **Rice** in risotto

Browning

Browning is a **change of colour** in food and can be either attractive or unattractive to the consumer. There are four basic causes of browning.

This is why grills are added to microwave ovens to achieve the effect of non-enzymatic browning.

Antioxidants are grouped as E300s.

Non-enzymatic browning	This is the reaction of carbohydrate and a protein, producing flavour, aroma and a brown colour. The outside of meat does this during cooking and is generally considered appetising.
Caramelisation	This happens when sugar is heated above its melting point and turns brown, producing a pleasant flavour. Take care not to overheat it or it will turn black and be bitter to taste.
Enzymic browning	This occurs when some fruits, e.g. apples, bananas and pears, are cut and left exposed to the air, turning them brown as the oxygen reacts with them. It can be avoided by: ● heating ● adding an acid like lemon juice ● adding a hot syrup ● blanching
Dextrinisation	This is the browning that occurs when starch turns to dextrin during cooking, e.g. bread, toast and cakes.

Colouring

The colour of food can be **unappetising**, especially if it is not the colour expected. Ingredients will often change colour when processed or cooked, so **colouring agents** are added to enhance or return the colour to its original.

> **KEY POINT**
> These **additives are not liked by some people, as they are not necessary for nutrition or taste.** In some cases additives have been **linked to allergic reactions.**

- Remember to use **natural ingredients**, such as chopped parsley, twists of cucumber or lemon and tomato slices as colourful and attractive **garnishes**.

Emulsifying

- **Emulsifiers** allow other liquids such as fats and oils to **mix together** in an **emulsion** and produce a **smooth** product, e.g. egg in mayonnaise.
- **Beating** will assist this process as it increases the surface area of each component.

Flavouring

Flavour is a combination of **taste and smell**, senses which can be affected if someone has a cold or smokes. Flavour can be lost in processing and so flavourings are sometimes added to **restore or enhance** the taste.
There are three classifications.

Natural	Produced from natural products
Nature identical	Chemically produced to be a copy of the natural product
Artificial	Chemically produced to create a desired effect

Monosodium glutamate is a commonly used flavour enhancer.

Fortification

Fortification is the addition of vitamins and minerals, especially those lost by processing or storage.

Laminating

To avoid problems when two **different textures** are used together, another **layer** can be applied.

Moistening

Some foods will be naturally **dry** and therefore **less palatable**. The addition of liquids or ingredients containing moisture will alleviate this.

Preserving

Food can suffer **deterioration** caused by bacteria, fungi and moulds. **Preservatives** can increase the length of time food can be stored.

- Sugar in jam preserves the fruit.
- Vinegar will preserve the vegetables in pickles.
- Salt can preserve fish.

Setting

Eggs, gelatine and cornflour are all used to set jellies and cold sauces.

Shortening

Shortening is the addition of fats and oils to flour in order to achieve a **crisp and crumbly** texture.

Stabilising

Stabilisers, such as starch and egg white, keep an emulsion stable by forming a gel which makes it more difficult for the components to **separate**.

- The smaller the particles in the emulsion, the more likely they are to stay together.
- Without the use of a stabiliser, emulsions are likely to separate to some extent with time.

> Stabilisers are grouped as E400s.

Sweetening

Many foods such as fruit and honey can sweeten dishes, but the main one used is **sugar**. Artificial sweeteners such as **saccharin and aspartame** can also be used. They are:

- **sweeter** than sugar and so less is used, being referred to as **intense sweeteners**
- **lower in calories** and can be used to reduce sugar consumption on health grounds

> Once again, there are concerns about reactions and side effects.

Tenderising

Sometimes foods, such as some cuts of meat, need to be tenderised to make them more digestible.

- **Mechanical** tenderising simply beats the meat to break up tough sinews.
- **Marinating** is adding a liquid to soften the meat before cooking, by the action of enzymes.
- **Ageing** is the hanging of some meat, such as game birds, to allow natural enzymes to soften and loosen the fibres.

> Care must be taken not to leave food too long as it may spoil.

Texture

A dish having only one texture can be **dull** and so the addition of ingredients with different textures will add **variety**. A side salad is a good example of providing a contrast to a pasta dish in terms of both texture and colour.

Thickening

Flours, starchy vegetables and bread can all be used to thicken soups, stews and sauces.

Food processing

AQA
EDEXCEL
OCR
WJEC
NICCEA

Primary and secondary processing

Primary processing is the preparation of raw materials to produce edible foodstuffs, e.g. washing and trimming fruit and vegetables, or their conversion into materials suitable for making edible foodstuffs e.g. milling wheat into flour.

Secondary processing is the conversion of the products of **primary processing** into edible foodstuffs, e.g. flour into bread or cane into sugar.

KEY POINT — Primary processing does not materially change the properties of the material, whilst secondary processing will produce a product that is edible, usually by the combination of more than one component.

Cooking food

Food is cooked for these reasons.

- To make it **easier to eat**.
- To improve the **resulting flavour** of the food by combination and the addition of spices and herbs, etc.
- To maker it **safer** to eat, as heating to a high temperature kills harmful bacteria.
- To enable it to be **kept longer**.

The **heat** necessary for cooking is passed to the food in three ways.

Conduction	By placing the food in or on a substance that is hot, e.g. in a saucepan of boiling water or in a frying pan
Convection	By placing the food in an atmosphere that is heated, e.g. in an oven
Radiation	By placing the food in the direct rays of the heat source, e.g. under a grill

A **microwave oven** works by the food molecules, especially water and fat, absorbing microwave energy and vibrating. This generates heat and the food is therefore cooked by **conduction**.

Microwave ovens do not cook food evenly, so the food should be rotated regularly to avoid cold spots, and left to stand to allow conduction to take place.

KEY POINT — Microwave ovens are especially useful for defrosting, reheating and cooking high water or fat content foods quickly.

Food preservation

AQA
EDEXCEL
OCR
WJEC
NICCEA

There are many ways to preserve food.

Chemicals

Removing water **(dehydration)** from the food can destroy micro-organisms. Vinegar, sugar, alcohol, salt and smoke all contain chemicals that can do this.

A study of the industrial methods of food preservation, relevant to your product, would be an important topic for your coursework.

It is essential that refrigerator compartments are kept below 5°C and that chilled food is kept below this critical level during transportation.

Chilling

Chilling is used for produce such as salad vegetables that shouldn't be frozen, but are safe to be kept for short periods below **8°C**.

If cooked food is quickly chilled to **just above 0°C**, and then kept at a temperature **below 5°C**, the growth of the majority of bacteria is halted. The products are referred to as **cook chill** and need to be reheated before eating.

Sous-vide is a modified version of the cook chill process, in which raw or par-cooked ingredients are sealed under vacuum and pasteurised. This has the advantage of **extended shelf-life** and improved **quality**.

Drying

Drying is another method that relies on the removal of water from the food to halt the growth of micro-organisms. It is important that the foods are kept **dry during storage**, as they will readily **rehydrate** when water is added.

Spray drying	Used for liquids such as **milk and soup**. The liquid is sprayed as a fine mist into a hot air chamber where it turns to a powder.
Hot air beds	Used for solid foods such as **vegetables and pulses**. They are dried on the beds in a similar manner to the traditional method of leaving them in the sun to dry in hot, dry countries.
Accelerated freeze drying	Used for **frozen foods**, where the ice is driven off as water vapour in a vacuum at low pressure. As this process does not damage the food or spoil its flavour, it is suitable for **heat sensitive** foods.

Accelerated freeze drying can also be used for **instant coffee granules**.

Irradiation

This method will **kill most** micro-organisms and delays fruit and vegetable ripening.

- Toxins are not destroyed and there is the usual loss of a small amount of vitamins.
- However, its use as a method of preservation is still controversial and, although allowed by law in the UK, it is not widely used.

 KEY POINT   **Irradiated food must be clearly labelled as such by law.**

Fig 3.3 The radura symbol

Freezing

See Section 3.5 for details of the star marking system for frozen goods.

Once water is frozen, micro-organisms and enzymes become inactive, but are not killed. However, in order for food to keep its **shape and appearance**, especially after defrosting, it is important that the freezing process reduces the temperature from **0°C to −4°C** in less than **30 minutes** so that only small ice crystals develop.

Industrial methods of freezing include the following.

- **Plate freezing** for thin, flat foods such as burgers and fillets
- **Blast freezing** for bulkier goods such as vegetables, where very cold air is blasted around them
- **Fluidised bed freezing** is used for small items that need to remain loose after freezing, such as peas. The cold air is blown upwards, keeping the food moving as in a liquid.
- **Cryogenic or immersion freezing**, which uses liquid nitrogen. It is very effective for delicate fruits as it freezes almost immediately, but is expensive.

Remember, once frozen food has been defrosted, it should never be refrozen.

> **KEY POINT**
> It is essential that domestic freezers are kept below −18°C, whereas commercial freezers operate at −29°C.

Blanching

Blanching is immersion in boiling water for a short period before rapidly cooling. It also improves the quality of frozen vegetables by stopping the enzymes and **reducing vitamin loss**.

Heating

Pasteurisation is only useful as a short-term method.

As heat kills micro-organisms and enzymes, it is a useful means of preserving food.

- During **pasteurisation** the product, most commonly milk, is heated to **72°C** for a minimum of **15 seconds** and then cooled quickly to **10°C**.
- **Sterilisation** is extremely useful for food that is to be **bottled or canned**. The canning process involves placing the food in a can and then, in the case of meat and vegetables, sealing it before it is heated to **115°C for 30 minutes**.
- **Ultra heat treatment (UHT)** is used for milk needing to be stored for six months or more. The process, where a high temperature of over **130°C** is maintained for **1 second**, has little effect on flavour or nutritional value.

> **KEY POINT**
> Remember that 72°C is the minimum maintained temperature that food should be heated to throughout.

Modified atmosphere packing (MAP)

This process seals the food in an atmosphere of **carbon dioxide and nitrogen** gas, which almost completely replaces the oxygen and therefore slows down the growth of micro-organisms.

As with MAP, once the packet has been opened or the seal broken, the food should be treated as fresh food.

Vacuum packing

All the air is removed from the packaging **before sealing**, once again preventing the growth of micro-organisms.

3.4 *Product development*

LEARNING SUMMARY

After studying this section you should be able to:

● *understand why and how new products are conceived*
● *understand how they are tested*
● *understand how they are put into production*
● *understand the incorporation of new technologies*

Product generation

AQA
EDEXCEL
OCR
WJEC
NICCEA

The aim of the food industry is to satisfy demands based on consumer needs. These needs may vary from those of basic nutrition to those of pure enjoyment. Financial considerations will also play a major part in determining consumer demand.

The successful company will achieve this customer satisfaction by:

● maintaining the **consistent quality** of existing products that are healthy and pleasant to eat
● providing an increased and **varied range** of products
● developing **new methods**, systems and technologies that increase **efficiency** and reduce waste
● reacting quickly to changes in consumer buying habits and **lifestyles**
● promoting brand names and **brand loyalty**

New ideas for products will be generated for a number of reasons.

● Reduced market share of existing products, because of new products from competitors
● Changes brought about by legislation or health related issues
● Changes in available production technology
● New market opportunities due to special events or minority needs

Feasibility studies

Research is undertaken to establish a **product specification** and to consider and establish whether:

Market research will often be used in the form of questionnaires or focus groups.

● there is a **genuine market** for such a product
● the product can be **produced effectively**
● the potential selling price will generate a **profit** and cover the costs of development, launching and production

Recipe formulation

Having created the product specification, and if the initial screening suggests that it is potentially successful, then the development team will formulate a recipe so that tests can be carried out. At this stage, the recipe is developed on a domestic scale – **one-off or small batch** – and evaluated by the company's own development team who take account of criteria such as main ingredient, type of dish, method of cooking, storage and size of portion.

See Chapter 1, p. 11 for details of different scales of production.

Each test sample must come with a complete reference.

● An exact list of ingredients and their relative proportions
● Processing methods, including differences required for industrial production
● Serving details, including links with other company products

Sensory testing

AQA
EDEXCEL
OCR
WJEC
NICCEA

The development team will include a group of **tasters** who have experience in testing new products for these features.

- Taste and flavour
- Consistency and texture (mouth feel)
- Appearance

> **KEY POINT**
>
> **These qualities are termed organoleptic.**

Some of theses samples will now be rejected for a number of reasons, whilst others will have suggestions made for **modifications**. This process continues until there is an agreed group of samples to take forward for further market research or **field trials**.

There are three basic types of tests: evaluation tests, descriptive tests and discrimination tests.

Evaluation tests

A panel of targeted consumers is simply asked to indicate, on a predetermined scale, how much it likes or dislikes a product. The tasting often takes place under coloured lights so that the colour of the food will not affect the response to other sensory descriptors.

This will only give an idea as to the potential success of a product. More sophisticated sensory tests need to be carried out to refine the product.

Descriptive tests

- The panel is asked to place a number of samples **in order** with reference to a certain characteristic. Symbols are often used to designate each sample, to avoid any subconscious association by the taster with the order of letters or numbers.

> **KEY POINT**
>
> **This is known as a ranking test.**

- The panel is asked to **score** the sample on a numerical scale with reference to a number of characteristics or **sensory descriptors**.

> **KEY POINT**
>
> **This is known as a rating test.**

The results of a rating test can be plotted on a **star profile**, with the arms of the star each representing a different sensory descriptor. The star profiles can then be compared easily, especially by overlaying, and the criteria for further adjustments formulated.

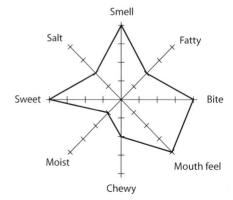

Fig 3.4 A star profile used in testing

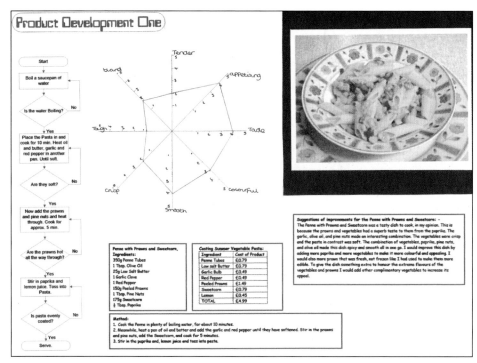

Fig 3.5 Star profile used in coursework

Discrimination tests

Paired comparison tests	Used to test two samples to see if the taster can tell the **difference** between them with respect to a specific characteristic.
Triangle tests	Are used to test small differences in samples, often in an attempt to replicate another product. The taster is asked to find the '**odd one out**', having been told that two out of the three samples are identical.
Taste threshold tests	Are used to determine how **little** of a substance needs to be added to achieve the desired result, e.g. salt.

Production

AQA
EDEXCEL
OCR
WJEC
NICCEA

Technical development

Once the testing has generated a preferred recipe, the formulation is scaled up to allow a **trial production** run.

 KEY POINT This will check that the recipe can be made, with the same results, using the actual production facilities available and the ingredients in the commercially required quantities.

Should any alterations need to be made to any of the working properties, a further test will be carried out to determine whether there has been any change to the flavour. Further organoleptic tests will be carried out to verify the results of the **batch production** samples.

During the factory technical development trials, the process is checked at all the **critical control points** to prove the system is working to specification and to carry out **hazard identification** to control risks (**risk assessment**) – especially with regard to contamination.

Critical control points

The process is monitored regularly at these points and **feedback** given to enable corrective action to be taken. The efficiency of this **closed loop** system can be critical to the profitability of a product – especially if there is a **high-risk** process involved.

>
> **KEY POINT**
>
> Many food contracts are on a 'just-in-time' basis because of the storage and shelf life requirements. The loss of a batch could seriously jeopardise the contract and the reputation of a company – as could a case of food poisoning.

See Chapter 1, p. 12 for further details on systems and control.

Control points will include the following.

Raw materials	These are checked for condition and arrangements made for suitable storage and preparation when potential contamination hazards can be identified.
Recipe formulation	Checks are made to verify that the correct ingredients are being used and that the quantities are within the agreed tolerances.
Cooking	Temperature and time are constantly monitored – usually automatically.
Cooling	The time taken to cool prior to packing must be strictly monitored and controlled.
Handling and packaging	Hygiene regimes must be maintained at this stage.

> **KEY POINT**
>
> The system of analysing the risks involved at the various stages of the production process, and then monitoring closely these areas during production, is a safety system known as hazard analysis and critical control points (HACCP).

Marketing

While the technical development has been in progress, the packaging has been designed in line with the **company image** and a **marketing strategy** developed so that production can start and the product **launched** with maximum market penetration.

New technologies

AQA
EDEXCEL
OCR
WJEC
NICCEA

Not all technological advances are accepted without question, even if the desired outcome is for the benefit of humanity. One such is **genetic modification (GM)**, a development of **biotechnology**.

The aim is to discover the genes that are responsible for unwanted characteristics, e.g. softening in fruit, and replace them with genes more adapted to the requirements – often from other species – or 'switch them off' altogether.

>
> **KEY POINT**
>
> It should be realised that food producers have been doing this 'naturally' for generations through selective breeding and hybridisation. It is a fear of the unknown, and the effect on the environment and its ecosystems, that is the cause of concern.

3.5 Packaging

LEARNING SUMMARY

After studying this section you should be able to:

● *recognise the importance of labelling*
● *discuss the functional requirements of packaging*

Food labelling

AQA
EDEXCEL
OCR
WJEC
NICCEA

> An investigation of product labels would be a useful component in your coursework.

The labels that are put on food packages generally provide the information required by the Food Safety Act and the Food Labelling Regulations of 1995. They also include other information that the manufacturer feels will promote its product and are therefore useful for the following reasons:

● Product recognition, promotion and comparison with other products
● Ingredients and dietary information
● Storage and preparation information

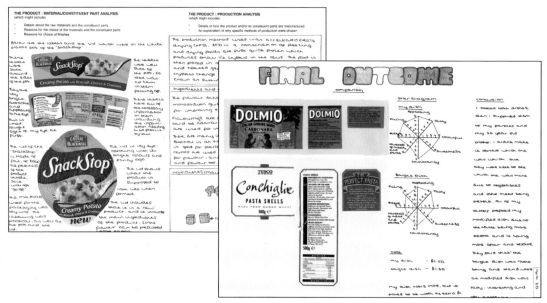

Fig 3.6 Product labels used in coursework

Product name

The name must:

● describe the product
● be a name generally used in the region where it is sold
● be legally established

A **subtitle** must be added to a name if it fails adequately to describe the food and must also indicate whether the food has undergone any process such as 'smoking'.

> **KEY POINT**
> If a picture is included on the packet, it must not indicate anything that would mislead the consumer, such as depicting real fruit in a dish that only contains fruit flavouring.

Name and address

The **country of origin** and the full name and address of the **manufacturer or packer** must be included.

List of ingredients

The type of additive should also be indicated, e.g. preservative, before the name.

- The ingredients used in the manufacture of the food should be listed in **descending order of weight**.
- Any **additives** used are also included in this list, using their specific names and/or the approved **E number**.
- The **average** weight or volume of the product that the package should contain must also be included.

> **KEY POINT** Some manufacturers also indicate the percentage of the total weight of some ingredients. This method of labelling is known as 'quantitative ingredient declaration', or QUID.

User and storage instructions

Cooking or reheating instructions are especially useful on products such as ready meals, as are instructions on how to **store** the product before and after opening the packet or container.

Shelf life

This refers to how long the product can be kept before it is opened and used. After this, the storage instructions will have to be followed to avoid spoilage. The date codes indicating 'shelf life' come in a number of forms.

Use by	Used for highly perishable foods that are usually only chilled and are thus more liable to become a health risk. Once the date shown has passed, the food is likely to deteriorate rapidly.
Best before	Used for longer-lasting items having less potential for microbiological spoiling. It is given in the form of the day and the month for items with a shelf life of less than three months. The year is added for longer-lasting items.
Best before end	Used with items that can last over 18 months. The month and year, or even just the year, need only be given.

Retailers will often mark down the price of items approaching their 'display until' date to promote their sale.

Other information that is often given, but not required by the regulations, is **'display until'**. This date is often set a few days before any other form of date coding in order to give the customer some leeway after purchasing.

>
> **KEY POINT** Fresh fruit and vegetables will often have this form of labelling, as they are not legally required to have other forms. It therefore also serves the purpose of building customer trust in the retailer.

Star marking	Consume within
*	1 week of purchase
**	1 month of purchase
***	3 months of purchase

Frozen foods benefit from having the 'star marking' system which indicates how long the product may be stored in a correctly functioning freezer cabinet or compartment.

Nutritional information

Unless the food company is making a specific claim about the nutritional value of a product, it is not required to include nutritional information on the label, although it is now common practice to do so as the consumer has become more **health conscious**.

> **KEY POINT**
> This information is usually given in terms of a **fixed weight or volume** so that direct comparisons can be made, e.g. per 100 gm, or in terms of a fixed **portion** of the product.

In your coursework you could compare a range of packaged foods by disassembling them from the information contained on the labels. This could include comparisons of ingredients, quantities and nutrition.

Other information

Depending on the space available, and what the company wants to achieve, other information can be included.

- Batch number – to enable production details to be investigated
- Coupons – money-off vouchers and free gifts of all kinds are used as marketing strategies
- Price – usually present only if the company is promoting the product by subsidising the retailer
- Guarantee – as these don't affect your statutory rights, they are another aid to gaining customer confidence
- Banners and flashes – used to bring attention to a specific feature e.g. 'New'
- Symbols – a variety of recognised symbols as illustrated in Fig 3.7

Indicates average weight of the contents, although the actual weight may vary slightly

Fig 3.7 Examples of the symbols used on labels

Functional requirements

AQA
EDEXCEL
OCR
WJEC
NICCEA

The packaging that food comes in has a number of practical advantages.

- Ease of transportation and handling
- Ease of storage and display
- Protects from damage and contamination
- Increases shelf life
- Convenient for labelling and other information
- Can help create and maintain a brand image, generating customer loyalty

> **KEY POINT**
> As packaging constitutes over 25% of household waste, you should concentrate on ways of **reducing packaging** by limiting its use to what is essential, re-using containers and recycling much of the rest.

Types of packaging

Primary packaging	The actual package or container that holds the product, e.g. bottle, can, wrapper
Secondary packaging	The packaging that either supports the primary packaging, e.g. the card cereal box that holds the inner bag, or the packaging that groups a number of products together, e.g. the box holding a number of drinks cans
Transit packaging	Any packaging used primarily for transportation, e.g. wrapped pallets

Materials for packaging

Glass	• Although comparatively heavy and easily broken, it has the advantage of being relatively cheap, re-usable and readily recycled • Extremely good for 'wet' products, maintaining its shape and being heat resistant • Can be printed on, though a label is usually added • Forms the greatest proportion of waste packaging
Metal	• Aluminium and steel are the next most common materials used in packaging, being both rigid and resistant to damage • Good barrier to liquid and gas • High energy costs to produce and requires coatings to prevent reaction with contents • Is readily recyclable and easily printed on although labels are still commonly used • Cannot be used in a microwave
Plastic	• Available in a rigid form as well as a film • Not as strong as glass and metal, but much lighter • Extremely suitable for all types of shrink and blister wrapping, and can be moulded into intricate shapes • Available in a wide range of colours and textures, and is readily printed on • Provides a good protection against moisture • It is not readily recycled, although progress is being made to make it economically viable to do so
Paper and board	• Light and reasonably strong, it often forms the outer packaging to improve features of other materials used • Readily printed and recycled • Poor protection against moisture • Chemicals used in its production could contaminate some foods if direct contact made

See Chapters 4, p. 73 for more detail on package design and Chapter 5, p. 91 for more information on materials – especially plastics.

Tamper-proof packaging

Many products' packaging must remain **airtight and sealed** until needed for health reasons, e.g. MAP packaging, bottled baby foods. There have also been cases of **malicious tampering**. It is therefore important that there is some way of indicating that the packaging has maintained its original state. Some of the ways that have been developed include the following.

The term 'tamper proof' is misleading – the systems noted can only give an indication that something may be wrong.

- **Shrink wraps** that cover the opening elements
- **Paper sealing strips and plastic break collars** that give a visual indication if they are torn
- **Vacuum button pops** that indicate if air has penetrated the closure.

Graphic products

The following topics are covered in this chapter:

- **Graphic presentation techniques**
- **Formal drawing techniques**
- **Visual representations**
- **Industrial practices**

4.1 Graphic presentation techniques

LEARNING SUMMARY

After studying this section you should be able to:

- **select and use the most appropriate graphic media**
- **recognise different styles of graphic illustration**
- **use colour and rendering techniques to enhance presentations**
- **distinguish between the various types of pictorial drawing methods**

Tools and equipment

AQA
EDEXCEL
OCR
WJEC
NICCEA

Using softer pencils requires care, as they can be messy. It is advisable to use a fixative if they are used a lot.

It is important that fine liners are used vertically to reduce wear and give a true thickness line.

You will be expected to acquire the **skills** to use effectively the following **graphic media**, judging when it is most **appropriate** to do so.

- **Drawing board** – preferably A3 size, T square, 60°/30° and 45° set squares, 300 mm rule, protractor, compass and dividers
- **Lead pencils**
  - 2B, HB for shading/toning
  - 3H for lining in, 6H for construction
- **Coloured pencils**

The quality of the pencils you use will directly affect the quality of your work. A **good quality** set will also **last longer** if looked after.

- **Fine liners** – 0.1, 0.3, 0.5 and 0.7 mm
- **Pencil sharpener, eraser and eraser shield**
- **Templates**

KEY POINT This is the minimum required – you may wish to use more to enhance your work further.

If available and relevant, the following could be used.

It is advisable not to use felt tip pens as they give a poor quality finish.

- **Airbrush, inks, paints, colour wash and gouache**
- **Marker pens**
- **Transfer letters**
- **Flexi and French curves**
- **Photocopier**
- **Light box**

- You will be expected to effectively use **computer software** for **CAD/CAM** and **DTP**, including such pieces of equipment as **template cutters**.

> **KEY POINT**
>
> However, it is most important that you understand how such applications are used in industry, including the advantages and disadvantages of using them, for the examination.

Types of graphic illustration

AQA
EDEXCEL
OCR
WJEC
NICCEA

This is an important skill to master and requires regular practice if it is to be used effectively. The extra effort will be rewarded.

You need to know the difference between the **four basic types** of illustration. They are each appropriate for use at different stages of your coursework as they differ in terms of **time taken** and their **accuracy**. The examiner will also use these terms, and you may **lose both marks and time** if you fail to understand the difference.

Freehand sketch

This method relies purely on you using the **pencil and/or fine liner** on its own to produce the illustration. It is therefore:

- the **quickest** method
- often used for **initial ideas or quick explanations**
- **lacking** in **accuracy** and **detail**

Sketch

This is the same as freehand sketch, but you '**crate out**' the basic shape first by using a rule or other equipment to draw a box first to which the shape is slowly added. The **final detail** is added **freehand**. This will:

- take more **time**
- give a more **satisfactory** end result

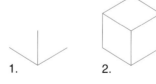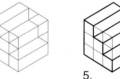

1. 2. 3. 4. 5.

Fig 4.1 The stages in 'crating out'

> **KEY POINT**
>
> Unless you find it impossible to do otherwise, it is advisable not to use this method in an examination if the examiner has asked you to produce a sketch, as it will lose you valuable time.

Drawing

Once you start to **develop** your ideas more fully, it is often desirable to produce a more accurate looking illustration. Drawings will:

- use **instruments** much more
- endeavour to maintain **proportion**

However, to save time, they will not:

- require complete accuracy in terms of **dimensions, angles**, etc

> **KEY POINT**
>
> All of the above three methods may be rendered as required to enhance the detail – see Rendering on page 63.

Construction

This is usually used for finished or **working drawings**. These drawings will:

- use **drawing instruments** at all times
- obey **conventions** and be dimensionally accurate
- be unlikely to be **rendered**
- take **more time** to produce

Enhancing the presentation

Although **single line** drawings are used extensively, it is more common to enhance illustrations with the use of **colour and rendering** techniques.

Colour

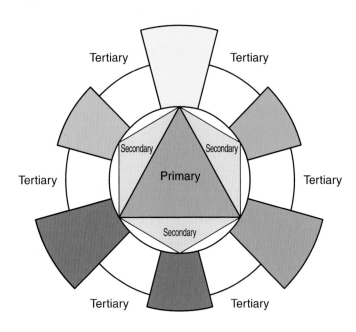

Fig 4.2 The colour wheel

The use of colour is extremely important in design, as it provides **further information** about our **surroundings** and has a direct effect on the way we **feel** about things and how we **react** to them.

The **primary colours** are **red, blue and yellow**, and cannot be made by mixing together any other colours.

Mixing any **two** of the **primary colours** together makes the **secondary colours**.

Mixing a **primary colour** with its **adjacent secondary colour** produces a **tertiary colour**.

Colours on **opposite sides** of the wheel are known as **complementary colours** as they are **contrasting**, whilst those colours **close** to each other will create **harmony**.

The basic colour is known as the **hue**, whilst adding differing amounts of white and black will create **tone**.

> **KEY POINT**
>
> **You will have noticed that the three primary colours** – magenta, cyan and yellow – **are the three** process colours **used in** colour printing.

Effects of colour

Different **colours** combined with different **shapes** can create impressions of relative **size and weight**. Although different societies and religions use different colours to represent different ideas, it is generally acknowledged that certain colours are connected in the mind with specific **feelings and emotions**.

Rendering

Rendering is the application of graphic media such as pencil, colour, marker, ink, etc., in order to create a more **realistic** representation by giving an **impression** of:

- depth
- light and shade
- texture
- material finish

KEY POINT This is a Key Skill, especially for the coursework, but is also worth practising for the examination.
Your sketchbook is useful in this respect.

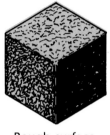  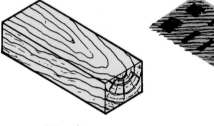 

Rough surface   High gloss   Wood   Textile

Fig 4.3 Applying rendering to achieve different material and finish effects

Three-dimensional (3D) or pictorial presentation

AQA
EDEXCEL
OCR
WJEC
NICCEA

Along with rendering, there are a number of **technical methods** of achieving a **three-dimensional** effect using a two-dimensional medium, i.e. paper.

KEY POINT It is important that you can use the most appropriate form to achieve the desired effect and know the difference between them for the examination.

Oblique

This is the **simplest form** of drawing in 3D, but is also the **least realistic**. However, it is often an effective method for the **less experienced** to use, especially when sketching.

- Start by drawing a true **'front view'** of the object on a **horizontal base**.
- All other **horizontal lines** are then drawn at **45°**.
- The **lengths** are then marked off, scaled down to **half size**.

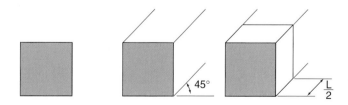

Fig 4.4 Stages in oblique drawing

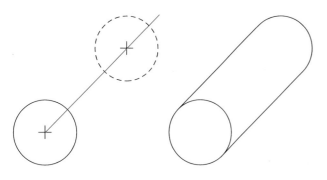

Fig 4.5 Oblique circles

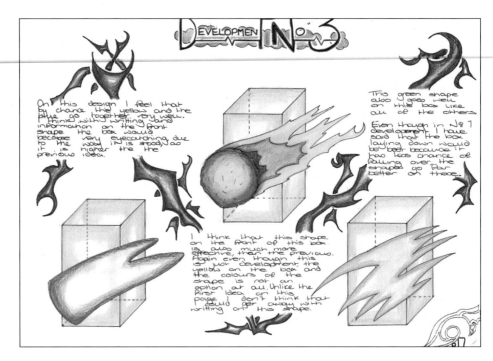

Fig 4.6 Example of oblique used in coursework

Isometric

Although more complicated in that it does not include a true face, this method produces a more realistic effect.

- Start by **crating out** using **30°** lines to represent **horizontal** lines, with **vertical** lines **remaining vertical**.
- Mark off **true scaled lengths** along the **vertical** and **30°** axes only.

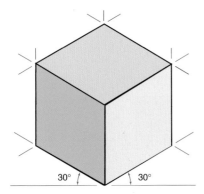

Fig 4.7 Crating out for an isometric drawing

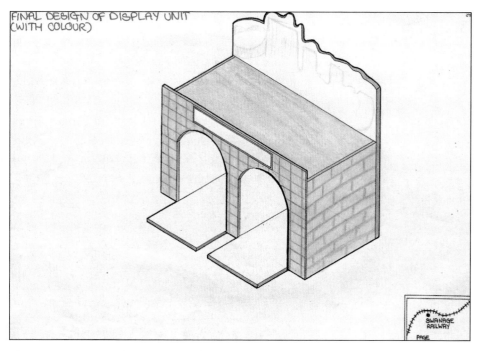

Fig 4.8 Example of isometric used in coursework

Isometric circles

Circles in isometric become **ellipses**, which have a **major and minor axis**. Use the line of the major axis, and the knowledge that a circle touches the **sides of a square** that encloses it at the **midpoints**, to locate the fixed points through which you can draw the ellipse.

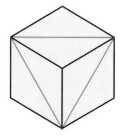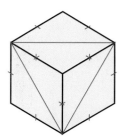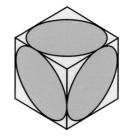

> **KEY POINT**
>
> An ellipse template is extremely useful for this in coursework, as you should be able to align the marks on the template once you have selected the appropriate size to suit your requirements.

Fig 4.9 Stages in drawing isometric circles

For more accurate **construction** of isometric circles, you should follow one of the methods shown in Fig 4.10.

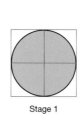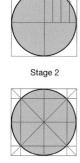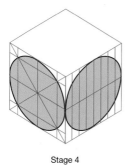

Stage 1

Stage 2

Stage 3

Stage 4

Fig 4.10 Constructing isometric circles

Planometric

In some design situations, such as kitchen design, the focus of the pictorial is the floor layout or plan. Planometric is used to give a three-dimensional impression of the finished design, leaving the important detail undistorted, unlike isometric.

● Draw the **plan view to scale**, but turned through **30°/60° or 45°/45°**.
● Project the **vertical lines** and complete the objects.
● Add finish detail, working from the front, but **leaving out** all, or part of, the **near walls** to give a more **'open' view**.

> **KEY POINT** Note **that in the case of 45° planometric, the vertical scale is** $\frac{3}{4}$.

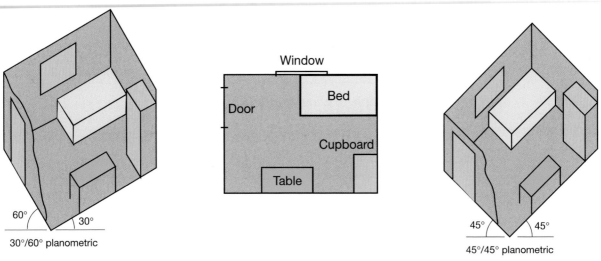

30°/60° planometric

45°/45° planometric

Fig 4.11 Planometric

In an examination it is advisable to use 30°/60° method if possible, to avoid having to use a $\frac{3}{4}$ scale on the vertical axis.

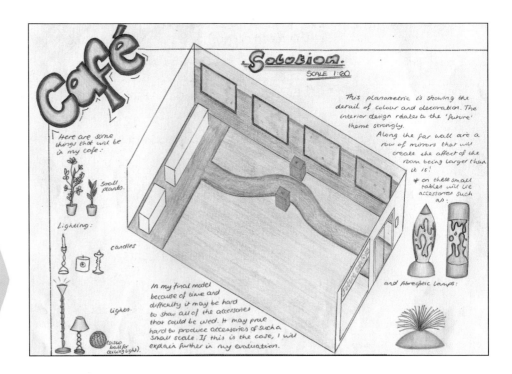

Fig 4.12 Example of planometric used in coursework

Single- or one-point perspective

You will have noticed that none of the previous methods gives a truly realistic effect because they take no account of **perspective**, where objects appear to get **smaller the further away** from the viewer they are (known as **foreshortening**). One-point perspective is based on the following principles.

- All **horizontal lines** moving away from the viewer meet at a **single point**, known as the **vanishing point (VP)**.
- The vanishing point is on the **horizon**, which is at **eye level**.

> This method is useful for curved objects as you start with a true front view.

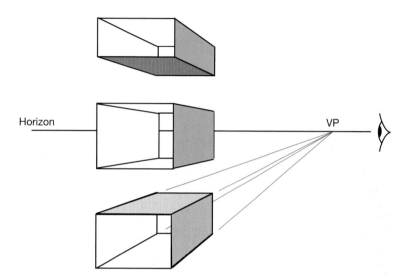

Fig 4.13 One-point perspective

Two-point perspective

Because the objects are still looked at 'front on' in one-point perspective, the result is still rather 'flat'. Two-point perspective makes the following assumptions.

- The receding horizontals to the **left and right** of the viewer meet at **two separate VPs**.
- These VPs are still on the horizon and at eye level.
- Everything is generated from the **leading edge** of the object.

> **KEY POINT**
> It is important to keep the two VPs at the edges of the drawing to avoid confusion.

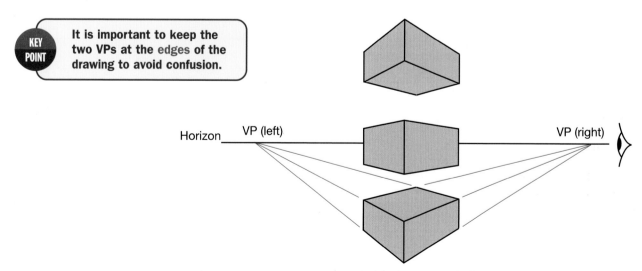

Fig 4.14 Two-point perspective

Estimated depth in perspective

The effects of the foreshortening already mentioned are most noticeable when a **perspective scene** contains elements that are **regularly spaced**. The following methods should allow you to produce this effect more accurately.

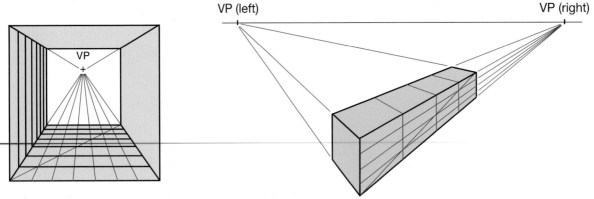
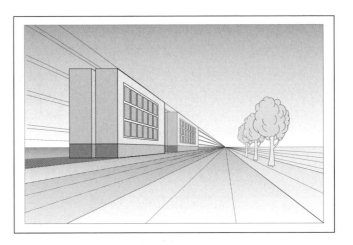

Fig 4.15 Estimated depth in perspective drawings

Fig 4.16 Example of one-point perspective used in coursework

Two-point perspective is a further improvement on isometric drawing. It should become natural to you with practice, improving the quality of your graphic communication.

Fig 4.17 Example of two-point perspective used in coursework

Perspective shadow

You will need to consider the effect of **shadow** caused by a direct **source of light** in many of your illustrations. You must consider the following when applying shadow in general.

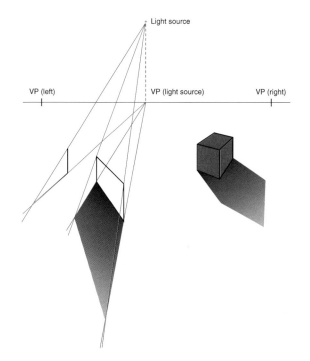

- The **position** and **intensity** of the light source
- How **directly** the light is falling on a **surface**
- How **reflective** is the surface

> **KEY POINT**
>
> Remember that shadow is not just dark or black areas. It can be achieved more effectively by blending different tones of the same colour.

However, in two-point perspective, shadows can be drawn quite accurately by assuming that if you follow your shadow back to the horizon, it will end directly under the light source, i.e. the sun.

Remember these points.

- Place the light source as **high** as possible, giving **shorter** shadows.
- Keep the light source to **one side** of the object.
- Treat the object as a series of flagpoles at each corner, connected at the top by a taut string!

Fig 4.18 Two-point perspective shadow

Thick and thin

When **time** and/or **printing costs** are critical factors in the production of an illustration, e.g. for a simple throwaway assembly/instruction sheet, and other methods of rendering are considered inappropriate, then the simplest way of creating a 3D effect is by the use of thick and thin lines. These are the rules to apply.

- If you can see **both surfaces** that form the line, it is drawn **thin**.
- If you can only see **one surface** that forms the line, it is drawn **thick**.

This will mean that:

- the **outside** or perimeter line will always be **thick**
- the thickness of **curved lines** around circular objects will **change along their length**

This is where two fine lines can be used to good effect, e.g. 0.3 and 0.7 mm.

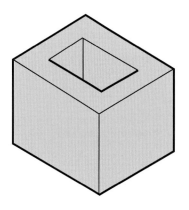
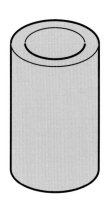

> **KEY POINT**
>
> A thick line is drawn approximately twice the thickness of a thin line.

Fig 4.19 Thick and thin lines applied to isometric drawings

4.2 Formal drawings

LEARNING SUMMARY

After studying this section you should be able to:

- *produce drawings in third-angle orthographic projection, including sectioning*
- *understand the line types to be used*
- *use scale and dimensioning*
- *produce exploded views and surface developments*

Third-angle orthographic projection

AQA
EDEXCEL
OCR
WJEC
NICCEA

This is the internationally recognised form for presenting technical drawings, which contains all the required information accurately **drawn to scale** so that others can manufacture the object. They are commonly known as **working drawings**, because they will sometimes also contain details of joints, materials, construction methods, etc.

KEY POINT

The symbol for third angle is: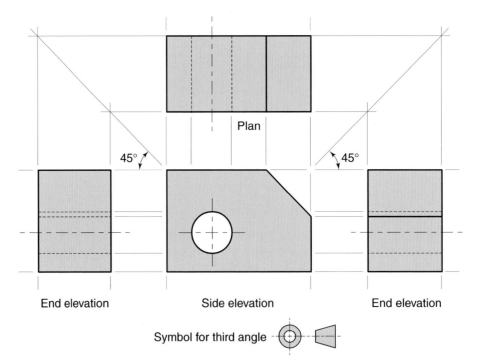

This **must** appear on the drawing.

Although four views are shown here, three are usually sufficient – especially in an examination.

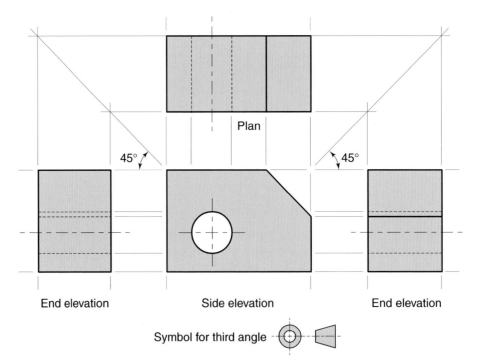

Plan

45°   45°

End elevation   Side elevation   End elevation

Symbol for third angle

Fig 4.20 Construction method for third-angle orthographic projection

KEY POINT

To remember the relative positions of the views – 'draw what you see from where you see it'!

Section views

AQA
EDEXCEL
OCR
WJEC
NICCEA

In some cases, the standard orthographic views – even with hidden detail, or because of it – will not clearly indicate the true shape or construction details of an object. A section view will assist this.

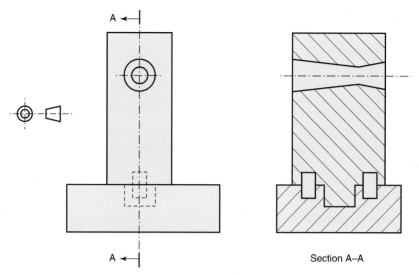

Fig 4.21 Sectioned view

Remember these points.

- Choose a section line that **'exposes' the unclear detail**.
- 'Cut' the object down the section line, removing the part **behind the arrows** and drawing what remains as usual.
- **Cross hatch** anything that has been **'cut'**, using different styles for different pieces.
- **Do not hatch** thin items such as **webs, nuts, bolts, screws, dowel,** etc.
- **Label** the section line and the view, ensuring the arrows conform to third-angle.

> **KEY POINT**   Do not **show** hidden detail **in section views.**

Line types

AQA
EDEXCEL
OCR
WJEC
NICCEA

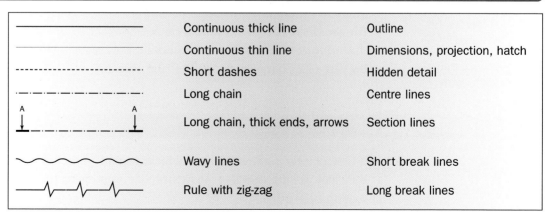

Line	Name	Use
———————	Continuous thick line	Outline
———————	Continuous thin line	Dimensions, projection, hatch
- - - - - - -	Short dashes	Hidden detail
— · — · — · —	Long chain	Centre lines
A — · — · — A	Long chain, thick ends, arrows	Section lines
∿∿∿∿∿	Wavy lines	Short break lines
—⅄—⅄—	Rule with zig-zag	Long break lines

Fig 4.22 Line types for working drawings

Scale

All orthographic drawings **must be drawn to a scale**, which is noted in the **title block** or elsewhere on the drawing.

A scale is a **ratio** and therefore has **no units**. Convention is that the scale of a drawing is the ratio between the drawing and the object, not the other way round.

Therefore a scale of:

- **1 : 2** indicates that the drawing is **half the size** of the object
- **1 : 1** indicates that the drawing is **full size**
- **3 : 1** indicates that the drawing is **three times the size** of the object

> **KEY POINT** Any dimensions added to a drawing, whatever the scale, will be the true dimensions – it is only the drawing that is scaled.

Dimensioning

Remember these points.

- Dimension **limit lines** should stop short of the object by **2 mm**.
- **Arrowheads** should be **long and thin** and not dominate the drawing.

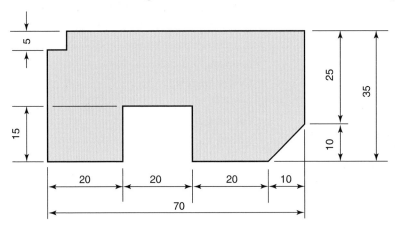

Fig 4.23 Dimensioning

Dimensioning needs to be thought about and can spoil an otherwise effective drawing. You are advised to refer to PD7308, a booklet issued by the British Standards Institute.

- Dimensions are written **above** the dimension lines and usually in the **centre**.
- **Vertical dimensions** are read from the **right-hand side** (having turned the drawing clockwise through 90°).
- The **smallest** dimension should be put **nearest** to the object.
- Use **millimetres** at all times and note it only once in the title block.
- **Do not** place dimensions '**inside**' the object.

Exploded views

This type of view is extremely useful for **assembly and instruction leaflets** as they indicate the **relative positions** of components without them obscuring each other.

Once again, the most common form of drawing used is isometric.

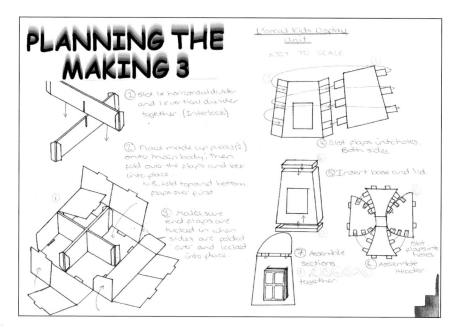

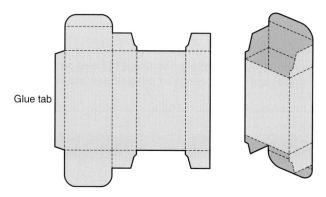

These are the basic rules.

- 'Explode' from the **middle out**, moving the components along the **30° and 90° axes** until they occupy their 'own space'.
- Use **dotted lines** to clarify the **relative locations** of some components if doubt could exist.

Fig 4.24 Example of exploded drawing used in coursework

Surface developments

AQA
EDEXCEL
OCR
WJEC
NICCEA

> You can understand more clearly how developments are used to create various types of packaging by disassembling them.

Many containers and packages are formed from a **single sheet** of material by cutting, scoring, folding, interlocking and gluing of tabs. The **flat forms** of such containers are called the **surface developments** (sometimes also known as **nets**).

The design of such developments is a critical process in the **packaging industry**, as is the way they are arranged on the raw sheet material in order **to reduce waste**.

KEY POINT When shapes naturally fit together in a regular pattern, it is known as tessellation.

Glue tab

Fig 4.25 Surface development of a container

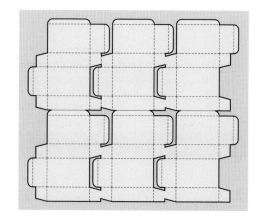

Fig 4.26 Developments arranged to reduce waste

4.3 *Visual representations*

Graphs and charts

AQA
EDEXCEL
OCR
WJEC
NICCEA

You often need to display the **results of surveys and questionnaires** in an easily accessible form. Graphs, charts and tables allow you to show your graphic communication skills to the full. You should consider the most appropriate use of:

● tables

● 2D and 3D bar charts, pie charts and line graphs

● pictographs

If you produce these by hand, use relevant graphic images to enhance/personalise the presentation.

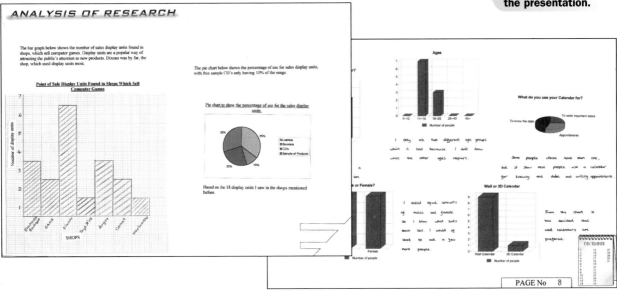

Fig 4.27 Examples of data handling used in coursework

KEY POINT

The production of these can easily be integrated by using ICT and data handling programs. Be sure to use the full range of possibilities available in terms of presentation styles.

Flow charts

AQA
EDEXCEL
OCR
WJEC
NICCEA

Flow charts are extremely useful when **planning any process** and can help determine the **'critical path'** – the sequence of events that will control the outcome of the process.

KEY POINT

A flow chart is not only a sequential list of operations, but it also contains feedback loops and control operations.

See Chapter 1, p.12 for more information on systems and control.

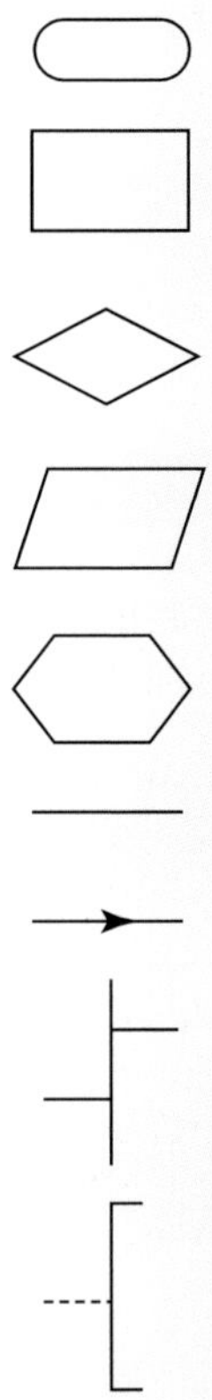

	Terminator	start, restart, interrupt, fault, stop
	Process	predefined, consisting of one or more operations specified already
	Decision	a question point in the process resulting in 'Yes' or 'No', and often used as a quality control point
	Input/output	shows the addition or removal of items or data
	Preparation	often follows a 'No' from a decision point
	Flow line	joins the symbols
	Flow line arrow	indicates direction of flow
	Junctions	should be well spaced and should not cross
	Annotation	broken line leading to explanatory notes

Fig 4.28 Flow chart symbols

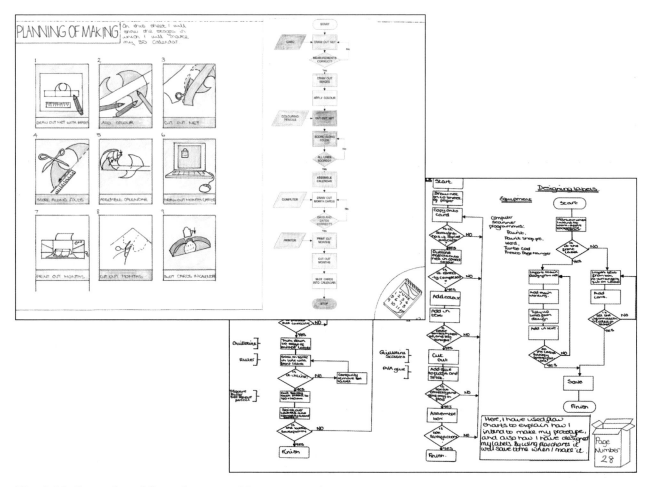

Fig 4.29 Examples of flow charts used in coursework

Schematic drawings

AQA
EDEXCEL
OCR
WJEC
NICCEA

Sequence diagrams

Sequence diagrams are largely graphic representations of processes designed to exemplify, and in many cases even take the place of, verbal or written instructions.

These are the main stages to consider when designing sequence diagrams.

- Decide on the **space** available for the complete illustration.
- **Split up** the whole operation into the **minimum number** of stages needed to illustrate the process, given the space available.
- Reduce each stage to its **basic components**.
- Produce the graphic **images** that clearly illustrate each stage.
- Add **notation** if necessary (or allowed!).

 KEY POINT Use existing known symbols **and such things as 'movement arrows' where possible.**

Fig 4.30 Sequence diagram for making a cup of tea

Schematic maps

Used to produce **simplified** route or location maps for publicity material or visitor guides to large attractions. The main aspects to consider when designing such maps are:

- the **removal** of unnecessary detail
- the **highlighting** of significant locations and/or facilities, often done by the use of an out of scale pictorial

Modelling

AQA
EDEXCEL
OCR
WJEC
NICCEA

 An important aspect of a model is scale, which may well need to include the weight of the model.

Modelling is an extremely useful form of communicating ideas. Models are used in all areas of Design and Technology, but will be the most likely way that you will cover the **material and manufacturing** requirements in Graphic Products.

Models can be used in a number of ways to:

- try out an **early idea** in the form of a **mock-up**
- test whether something works practically in the form of a **working model**
- give an **impression** of the finished object in the form of a **prototype**

 KEY POINT Remember that 'modelling' can also refer to simulations produced by more sophisticated computer programs, including 'wire-frame' modelling.

For further information on material properties, refer to Chapter 5, Resistant materials technology.

Modelling tests your **inventiveness**, as the main aim is to give an impression of the object – not to manufacture it. Therefore the **whole range** of materials is open to you, as long as the chosen ones **suit their purpose**. Some of the more useful modelling aids are discussed below.

Construction kits

The use of these kits will not be acceptable as the only evidence of making skills.

These are especially useful for developing mechanistic ideas and are used effectively as mock-ups.

Card

This is the most **readily available** material and is extremely useful as it is available in many grades, thicknesses, colours and finishes. It is easily cut and will accept many different adhesives and applied finishes. It is especially useful for **architectural models** and, of course, **packaging**.

 KEY POINT   Always **use a cutting board, safety rule and sharp scalpel.**

Foamboard

This is a specialist modelling board, excellent for **architectural models** and more rigid items such as **point of sale displays**. It is relatively **expensive** and requires special adhesives. However, its **rigidity** makes it easy to use and the card surfaces accept all forms of rendering.

Medium density fibreboard

A manufactured board that is much better for modelling, especially for the production of **prototypes**, than natural timber because it:

- has **no grain**, making it **easy** to cut and shape
- can be readily joined using wood adhesives such as **PVA**
- can produce an **excellent finish** using abrasives
- accepts applied finishes like **cellulose paint** sprays (especially with a primer/filler), producing an extremely realistic effect

 KEY POINT   As with all manufactured particle boards, cutting and abrading produces a lot of dust. Good extraction and the use of a dust mask are essential.

Jelutong

This is a natural hardwood, with even better working properties than MDF. However, it is expensive, and even the off-cuts should be retained for possible future use.

Plastics

Acrylic is often used as a modelling material because of its attractive finish and colours, but not always with great success. It is reasonably **difficult** to cut, form and join. It also **does not** readily accept applied finishes.

ABS is much more **useful** as it comes in thinner sheets and can easily be used with a **vacuum former** to produce inserts for packaging, for example.
Expanded polystyrene is readily available as a waste product from packaging, but is rather coarse and crumbly when worked, and is therefore only really suitable for the **early development** of ideas.
Styrofoam is a fine-celled version of expanded polystyrene, which can be **cut and formed readily**, and finished using fine abrasive papers. It is joined using PVA and can be painted using a water-based emulsion paint.

> **KEY POINT**
> Although polystyrenes can be cut easily using a hot-wire cutter, the fumes are toxic. Such a cutter should only be used where there are adequate forced air-extraction facilities, and never when the wire is red-hot.

Fig 4.31 Examples of models used in coursework

4.4 Industrial practices

> **LEARNING SUMMARY**
>
> After studying this section you should be able to:
> ● understand the properties of paper and board
> ● understand typography
> ● differentiate between the types of machine printing
> ● recognise different printing techniques

Paper and board

AQA
EDEXCEL
OCR
WJEC
NICCEA

The description of any paper or board should indicate:

● type
● colour and finish
● size
● grammage

Types of paper and board

There are several commonly used raw materials.

Mechanical wood pulp	• Uses **90%** of the debarked tree which is ground down mechanically, using the wood very efficiently • Contains **short fibres and impurities** which **weaken** it • Produces paper of a **poor quality** which **ages** quickly when exposed to **light** • Is used for **cheaper grades** of paper
Chemical wood pulp	• Uses **50%** of the debarked tree which is ground down **less finely** • Contains **longer fibres** and is **chemically treated** to remove impurities • Produces a stronger, **cleaner** paper that is **more expensive**
Recycled paper and board	• Is becoming more important as the pressure to **conserve raw materials** increases • Is used for **lower quality papers and board** because the **fibres lose strength** each time they are recycled and therefore must have new pulp added to preserve quality
Rag	• Uses cotton, linen, hemp, manila or jute, with **cotton and linen** producing the best quality • Have **long fibres** so the paper product is **strong and durable** • Are **expensive** to produce

These are the more commonly used papers.

Newsprint	• **Cheapest grade** of printing paper, made from mechanical wood pulp, including recycled pulp • Supplied in **reels or sheets** • Readily accepts the inks used in **newspaper production**
Mechanical printing	• **Superior newsprint**, containing more chemical wood pulp • Used for **cheap publications** and **writing paper**
Mechanical SC printing	• Used for **mass circulation** magazines and cheaper books
Woodfree printings	• Contain no mechanical wood pulp • Used for all kinds of **general printing**
Cartridge	• Used for **printing** as well as **drawing** • Produced in a range of **surface finishes**
Offset printings	• Used for **lithographic** printing
Coated papers	• Papers given different **surface treatments** for different printing processes
Banks and bonds	• Produced with a **matt surface** for typewriting and handwriting
Duplicators	• Have a **soft, matt** finish and are **slightly absorbent** • Good for **duplicating** processes, where a good impression is required
Cover papers	• A wide range of strongly **coloured** papers, usually thick, with good **folding and wearing** properties • Used for covers for **booklets and brochures** • When **laminated**, they produce a thicker paper known as **cover board**

 KEY POINT **A considerable number of the above papers are also available as** board **weights, which usually start at 200 g/m^2.**

The most commonly used boards are:

Pulp board	• Produced in various qualities and finishes
Index board	• Resembles pulp board, but with a **range of tints** and a **high machine finish** suitable for **printing and writing** • Used for **office records** and **card index** systems
Paste board	• **More rigid** than pulp board, having a **core lined** on both surfaces with a **higher quality paper** • **Two or more sheets** may be combined together to give greater thickness as **duplex or triplex boards**

Further information on paper can be obtained from:
The Pulp and Paper Information Centre
Papermakers House
Rivenhall Road
Westlea
Swindon
SN5 7BE

Finishes

KEY POINT
Apart from the **fibres** of a paper, over 30% of the weight is made up of **additives** which determine the quality, printability and general characteristics.

The main additives are:

Fillers	● Used to **fill the gaps** between the fibres to produce a **smoother surface**
Sizing	● Synthetic and resinous materials used to **bind the fibres and fillers** together ● **Retards** water penetration and the **spread of inks**
Colouring	● Dyestuffs and whitening agents used to produce the desired colour

Standard sizes

Although paper and board is still supplied in many different sizes, both trimmed and untrimmed, most papers are now supplied based on the **International Standards Organisation 'A', 'B' and 'C'** ranges.

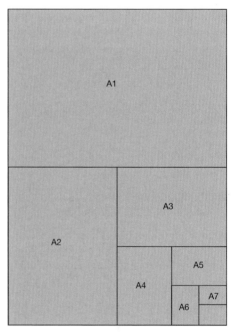

A range

A0	841	×	1189
A1	594	×	841
A2	420	×	594
A3	297	×	420
A4	210	×	297
A5	148	×	210
A6	105	×	148
A7	74	×	105
A8	52	×	74

KEY POINT
The most commonly used is the ISO A-sized range.

The basis for the A range is:

- a rectangle of area **one square metre** (1,000,000 mm²)
- the **sides** of which are in the **ratio 1 : √2** (1 : 1.414), i.e. **841 mm × 1189 mm**
- designated A0

KEY POINT
This ratio has the property that it is **maintained** when the longer side is halved or the shorter side is doubled.

Fig 4.32 The A range of papers obtained from the basic A0 sheet

Grammage

The standard international sizes are stocked in a range of standard grammages which indicate the weight in **grammes of one square metre** of the paper or board – **g/m².**

Typography

AQA
EDEXCEL
OCR
WJEC
NICCEA

Size of type

The size of any typeface is based on a method of measurement known as the **point system**. One point (1pt) measures 0.351 mm.

KEY POINT
The standard 12pt measurement is known as a **pica.**

The point size does not refer to the height of the letter, but to the space between the lines required to ensure that the **ascending letters** (b, d, f) do not touch **descending letters** (g, j, p). This is referred to as the **body size** of the type.

However, to achieve **impact** or improve **legibility**, it is sometimes desirable to increase the space between the lines – known as **leading**. This is achieved by putting a smaller point type on a larger point body. For example:

_ This is an 18pt type,
_ using an 18pt body.

_ This is an 18pt type,
_ using a 24pt body.

_ This is a 24pt type,
_ using a 24pt body

The sizes **up to 12pt** are traditionally known as **composition** sizes, whilst those from **14pt to 72pt** are known as **display** sizes. Larger sizes are known as **headlines**.

Type fonts

A font is the term given to an assortment of type characters of the **same size and design**. This will include the alphabet – upper and lower case, numbers, punctuation marks, reference marks and special signs.

> The English word for this was 'fount', pronounced font. The American spelling is now the norm.

Type characters

As letters are not all of the same width and configuration, to improve the visual appearance some adjustments are made.

- **Letterspacing** is the changes made to the standard space between characters to **remove gaps** created by **narrower** letters,

  e.g. Illustration
  Illustration

- **Kerning** is the adjustment of the space still further to allow one part of a letter to **extend over** the next,

  e.g. *fo*

- **Justification** is the adjustment of the spaces **between words** to allow the **edges** of the text to **line up**, or follow a predetermined line. This is most noticeable when both edges of the text are to be lined up.

Typefaces

A typeface is defined as a set of characters identifiable by design and available in a range of sizes. They are grouped into families and can be further changed by:

- *italicising*
- adjusting the **weight**
- adjusting the **width**

> Investigate the capabilities of your word processing package to take full advantage of these features.

KEY POINT
A common difference between typefaces is whether the design includes small strokes at the top and bottom of the main strokes, **which are known as serifs. Typefaces without these are called** sanserif.

Machine printing

AQA
EDEXCEL
OCR
WJEC
NICCEA

Offset lithography

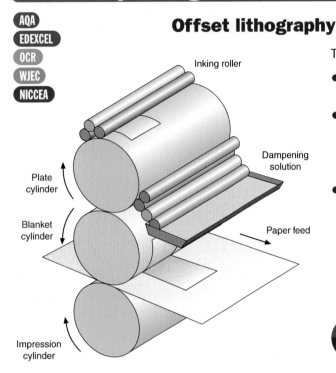

This is the basic process.

- The image is **transferred photographically** onto a **cylindrical plate** using ultraviolet light.
- The plate is washed in a chemical that **attracts** the ink to the **image surfaces**, and then wetted with a dampening solution that **repels** the ink from the **non-image surfaces**.
- The plate is then **inked**, rotated and pressed against a **blanket cylinder** that **transfers** the image onto the **material to be printed**. This blanket cylinder **prolongs the life** of the original plate and gives the process its name – **offset**.

> **KEY POINT**
> The material (or substrate) to be printed can be supplied in either sheets (sheet fed) or reels (web fed).

Fig 4.33 The basic arrangement for offset presses

Letterpress

Letterpress is now superseded by other means of rotary printing – especially offset lithography. It is a form of relief printing where:

- the printing surface is raised above the non-printing area
- the printing surface is then inked and pressed directly onto the printing medium
- the process can be carried out on a flatbed (or platen), flatbed cylinder or rotary press

The original must be a '**mirror image**' of the intended outcome.

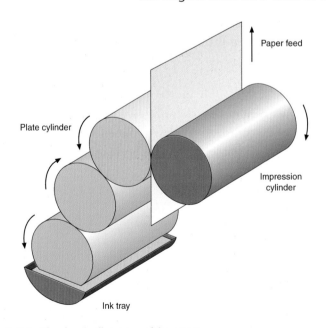

Flexography

This is now the dominant relief printing method, because of its **speed**.

- The plate and blanket cylinder are now **combined** with the image **directly engraved** onto the blanket cylinder material, which is a flexible but resilient rubber or photopolymer.
- **Fast drying** water-based or solvent inks are used onto virtually **any material**.

Fig 4.34 The basic flexographic press

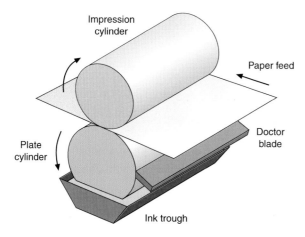

Fig 4.35 The basic gravure press

Gravure

This is essentially the opposite of relief printing, using the **intaglio** printing process, where:

- the printing areas are in **recess**, having been engraved onto the plate cylinder
- the **recesses** are filled with **ink**
- **surplus** ink is removed from the **non-printing** surfaces by a **doctor blade**
- the printing material is then **pressed** against the cylinder to **transfer** the ink onto it

Screen

> This is sometimes referred to as 'silk screen', because silk was the first material to be used.

Screen-printing is a form of **stencil printing**.

- The printing and non-printing areas are carried on a **mesh screen** usually made from **nylon, polyester** or **fine mesh stainless steel**.
- The **non-printing area** is '**blocked out**', leaving the image clear.
- The ink is then **forced through** the mesh directly onto the printing material.

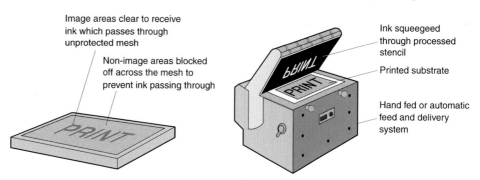

Fig 4.36 The basic screen-printing press

Printing method	Uses	Advantages/disadvantages	Economy
Lithography	Newspapers, cheaper magazines, books, business stationery, posters and packaging	• Widely available, producing good print quality • Reasonably short plate life, but readily replaced	• Best suited to larger print runs – 250,000 to 1 million • The choice for economical general printing
Letterpress	Private stationery and business cards	• Good quality outcome but slow • Becoming less widely available	• Economical for short runs only – 500 to 5,000 • Expensive for colour work
Flexography	Useful for packaging that uses materials other than paper, because of the quick-drying inks that can be used	• Lower energy costs, less waste and higher running speeds • Especially good for polythene food packaging	• High print runs required – over 250,000
Gravure	More expensive books and colour magazines, large print run colour work on poorer quality paper, food wrappings	• High quality work at high speed, but expensive to set up	• Large print runs needed – 500,000 to many millions
Screen	T-shirts and other fabrics, posters and shop display signs	• Relatively slow process, but adaptable. • Lacks fine detail, but can print onto many different surfaces	• Low cost option for short runs – 1 to a few hundred

Printing techniques

AQA
EDEXCEL
OCR
WJEC
NICCEA

Further information on printing processes can be obtained from:
British Printing Industries Federation
11 Bedford Row
London
WC1 R 4DX

Colour printing

- **Single-colour** work is usually black, but any colour can be used.
- **Four-colour process** work (or full colour printing) uses the three process colours – **yellow, magenta and cyan** – plus black.
- **Spot-colour** work uses a main colour (usually black) plus another colour or colours to **highlight** headings, borders, etc. This is much **cheaper** than four-colour work.

Colour separation

To allow a picture to be printed in colour, a camera or scanner is used to **separate** the **four** process colours and produce a plate for each colour. The colours are then **superimposed** on the printing material to create the full range of colours. Some more modern litho presses can have **extra units** for **special colours**, e.g. gold, and **finish coatings** (see below).

Registration

This is the process of **aligning** one colour exactly on top of the previous one to create the desired **finished** colour. The **registration marks** can be seen in the '**trim area**' of the sheets or web.

Colour control bars/strips

These are also printed in the **trim area** and are examples of what should be in the main body of the print. They therefore serve the purpose of **quality control**.

 KEY POINT Because printing presses can't print right to the edge of a sheet or web, there is an area known as the 'trim area' that is cut off as part of the finishing process.

Varnishes and coatings

To improve the **appearance** and **wearing and protective qualities** of printed matter, an additional varnish or coating can be added for the **same cost** as an **extra colour**. They can be applied all over or in one 'spot' and can be matt or glossy. In-line coating falls into three main types.

- **Varnishing** is similar to the application of a 'colourless ink' which gives good added moisture protection, but is **slow drying** and can **yellow with time**.
- **Aqueous coatings** give a hard surface, are **quick drying** without the yellowing, but **can distort** lower grammage papers.
- **UV varnishes** give the best finish in terms of **high gloss finish**, and **dry almost instantly**. However, they are **more expensive** than the other methods.

Laminating

Lamination is the application of a **clear plastic film** to the printed surface, achieving the best finish in terms of gloss and, especially, protection. It is much **more expensive** and is usually used for **cartons** and **'special'** presentation work such as reports.

> **KEY POINT** Paper and board can also be lined or laminated with a foil layer, **and this is often used in the** food industry.

Embossing

Embossing is produced on flatbed letterpresses or specialist machines.

Embossing – a **raised or depressed area** on the printed paper or board – is done to emphasise a feature by the use of relief, often combined with the use of another colour. It is the result of pressure applied on the printing material between a **hollow 'female' die** and a **'male'** counterpart.

> **KEY POINT** Embossing is commonly used on business cards, letterheads, brochure covers, high quality labels, folding box cartons, e.g. chocolate boxes.

Folding and creasing

To assist the folding of paper and board, a **'blind impression'** is made on a rotary machine – often at the same time as printing for thinner materials – using a **creasing rule** to **compress and stretch** the fibres at the point of the fold.

Die cutting

Irregular shapes for cartons, pop-up books and cards, etc., are cut and creased from sheets using **dies** made from **steel cutting and creasing rules** set in a **forme**. Die cutting is usually carried out on a letterpress or specialist machine.

Further information on packaging technology can be obtained from:
The Institute of Packaging
Sysonby Lodge
Nottingham Road
Melton Mowbray
Leicestershire
LE13 0NU

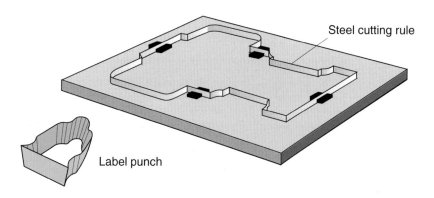

Steel cutting rule

Label punch

Fig 4.37 Simple die cutters

> **KEY POINT** Specialist **folder/gluer machines** then complete the process. These machines can often emboss and create windows and window patches to allow contents to be seen.

5 Resistant materials technology

The following topics are covered in this chapter:

- **Tools and equipment**
- **Materials**
- **Construction**
- **Surface finishes**
- **Manufacturing processes**
- **Quality**
- **Mechanisms**

5.1 Tools and equipment

LEARNING SUMMARY

After studying this section you should be able to select, name and use appropriately:

- **marking out tools**
- **clamping and holding tools**
- **cutting and shaping tools**
- **hole-making tools**
- **machine tools**

Marking-out tools

AQA
EDEXCEL
OCR
WJEC
NICCEA

- The basic purpose of marking out is to allow material to be cut accurately to size and shape. These are the best ways to mark out each material.

> Never use a pen on wood, as it will permanently stain the grain.

Wood	• A pencil for construction lines • A marking knife for lines that will be cut to
Metal	• Engineer's blue and a scriber for construction lines • A centre punch at intervals to make a witness line for lines that will be cut to
Plastics	• A wax crayon or felt tip pen for construction lines • A scriber for lines that will be cut to

- Always mark out from a prepared **true** or **face edge**.

KEY POINT Remember to mark the **waste side** of any line to be cut with **hatch lines**.

Tool	Tool in use	Use on	Notes	Tool	Tool in use	Use on	Notes
Scriber	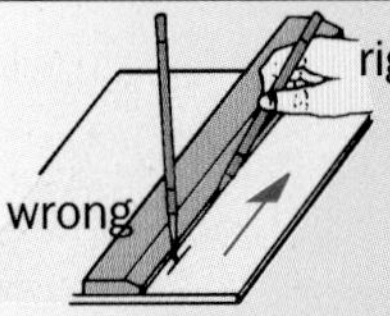 right / wrong	M P	Used to mark out on hard materials such as metals	Dividers		M P	For transferring measurements and marking out arcs and circles on metals and plastics
Engineer's square/ try-square	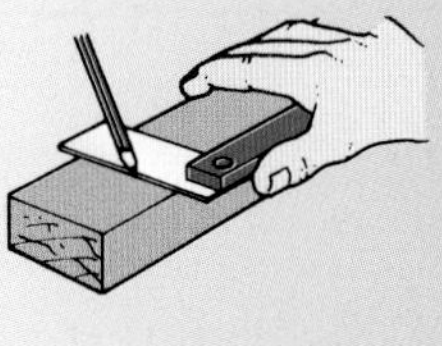	W M P	Marking lines at right angles to the edge of metals, plastics and timber. Also used to check right angles	Centre punch	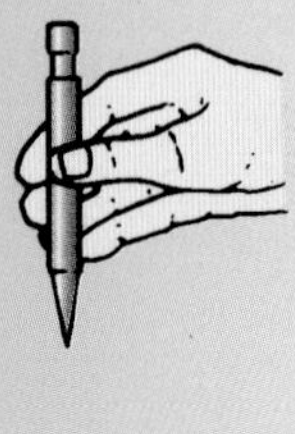	M P	Marking centres of holes before drilling or marking out with dividers

Fig 5.1 Marking-out tools

Key: W = wood, P = plastics, M = metals

Tool	Tool in use	Use on	Notes	Tool	Tool in use	Use on	Notes
Centre square	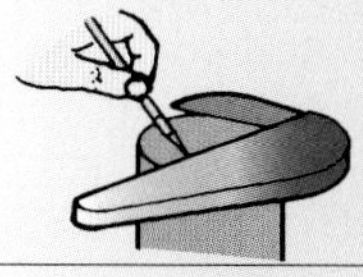	W M P	Used to find the centre of a round bar	Sliding bevel		W M P	Marking out and copying of angles
Micro-meter	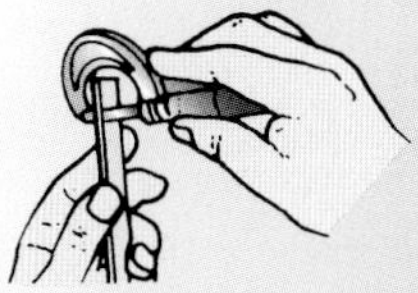	M P	Measuring and checking precise measurements. Cannot be used for internal measurements	Marking gauge	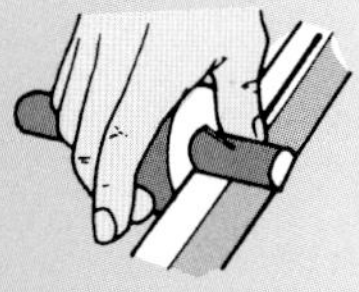	W	Marking a line parallel to an edge

Fig 5.1 Marking-out tools (continued)   **Key:** W = wood, P = plastics, M = metals

Clamping and holding tools

AQA
EDEXCEL
OCR
WJEC
NICCEA

Tool	Tool in use	Use on	Notes	Tool	Tool in use	Use on	Notes
G-cramp	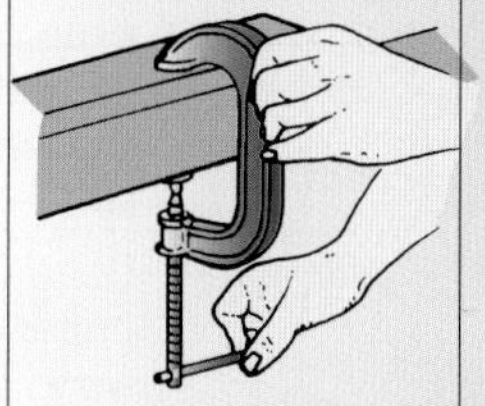	W M P	Used in almost all situations when two parts need to be held together for a short while, e.g. glueing up or drilling on the pillar drill	Machine vice		M P	Used to hold materials accurately and securely when drilling. The vice can either be hand held, bolted or cramped down
Tool-maker's cramp		M P	For holding work together when a G-cramp is not appropriate. Usually used on small pieces of work and rarely with timber	Hand vice	 scrap wood / drill table	M P	Used to hold sheet materials when drilling. Make sure the wing nut is tight and held securely in place
Sash cramp	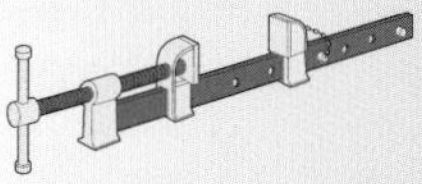	W M	Used to hold frames, carcasses and butt joints together during fabrication	Pliers	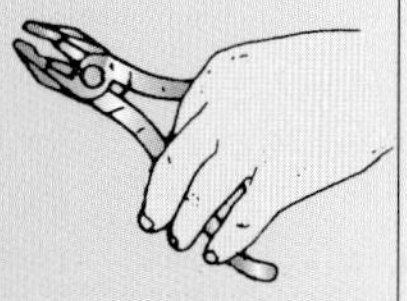	M P	Available in a variety of sizes. Used to hold small items when fingers would be too large or not strong enough. Do not use as a substitute for a spanner
Engineer's vice	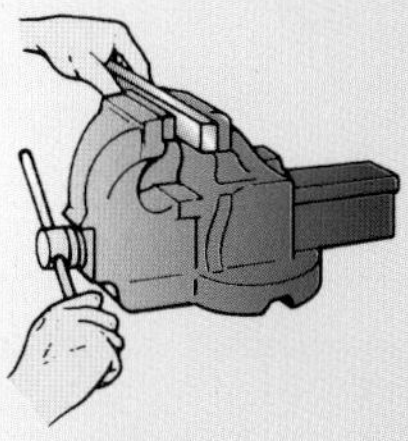	M P	Remember: the engineer's vice has jaws that may damage soft materials	Carpenter's vice	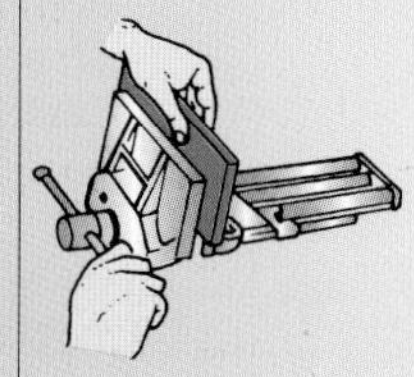	W	The carpenter's vice is better for holding large pieces of wood

Fig 5.2 Holding tools   **Key:** W = wood, P = plastics, M = metals

Cutting and shaping tools

Tool	Tool in use	Use on	Notes	Tool	Tool in use	Use on	Notes
Tenon saw	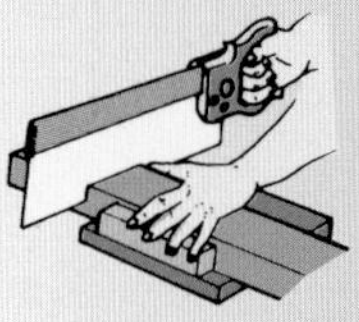	W	Used to cut relatively small pieces of wood and for sawing joints, such as tenons and halvings	Tin snips	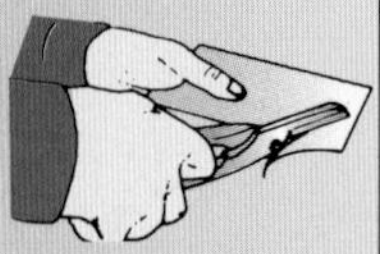	M P	Originally intended for cutting tin plate, but equally useful for sheet aluminium, copper, etc., as well as sheet PVC and polystyrene
Hand saw		W	Available in different lengths and with different numbers of teeth per cm. Used to saw larger sections of wood	Side cutters	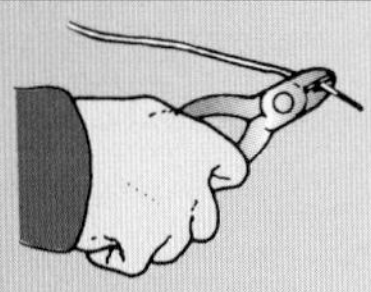	M P	Used to snip off surplus wire after soldering electrical components. Can be used to cut electrical wire, but they should not be used to cut hard (piano) wire
Coping saw		W P	Not ideal for straight cuts but good for curves and removing awkward shapes in wood and some plastics	Acrylic cutter	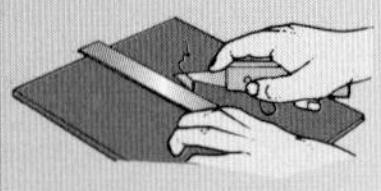	P	With this tool, the plastic (acrylic or polystyrene, for example) is scored and then broken along the line
Firmer chisel	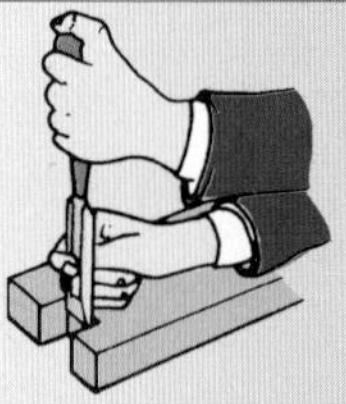	W	Generally used horizontally or vertically (as shown) for removing small amounts of timber accurately	Trimming knife	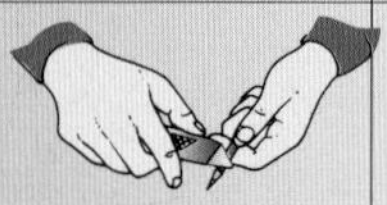	W P	Used for trimming paper, card, thin wood and plastics. It can also be used for accurate marking-out of wood
Jack/ smoothing plane	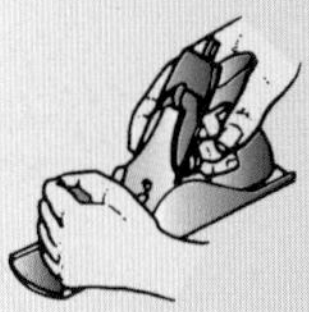	W P	Used to prepare wood accurately to size and for final cleaning, prior to using glasspaper	Surform	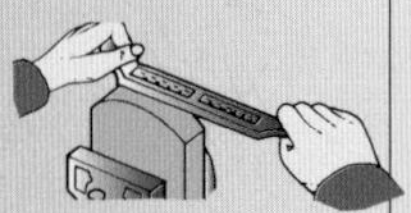	W P	A very versatile tool for free shaping of wood, soft plastics, plaster, etc. Used with care, it can shape plastic foams such as polyurethane
Taps and dies	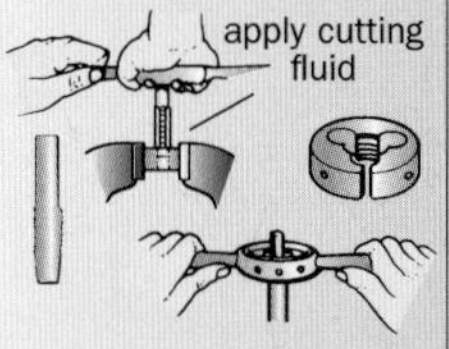 apply cutting fluid	M P	For cutting male and female screw threads. Sizes range from 3 mm up to 15 mm and larger	Guillo-tine/ bench shear	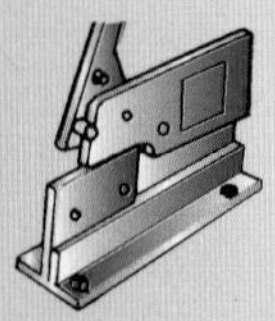	M P	For cutting sheet metal that cannot be cut with tin snips. It is usually locked closed when not in use, to avoid accidents
Files	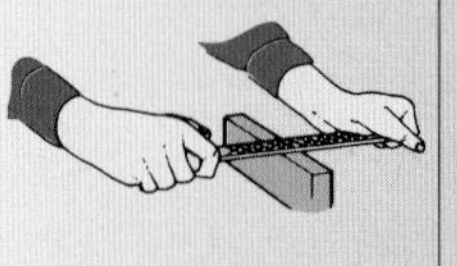	W M P	Available in a huge range of lengths, shapes and 'cuts'. Used on metals and plastics. If used on wood they tend to clog. Always make sure a handle is fitted	Hacksaw	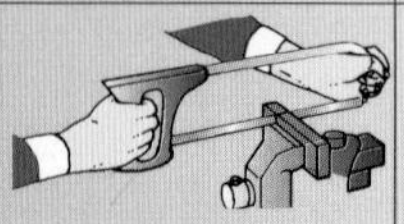	M P	For use on metals and plastics. The blade can be fitted sideways to allow cutting along the length of the material without the frame getting in the way

Fig 5.3 Cutting and shaping tools   **Key:** W = wood, P = plastics, M = metals

Hole-making tools

AQA
EDEXCEL
OCR
WJEC
NICCEA

KEY POINT The observation of safe working practices is important when using hand tools, as they can cause injuries that are equally as serious as those caused by machine tools.

Tool	Tool in use	Use on	Notes	Tool	Tool in use	Use on	Notes
Hand drill/ twist drills	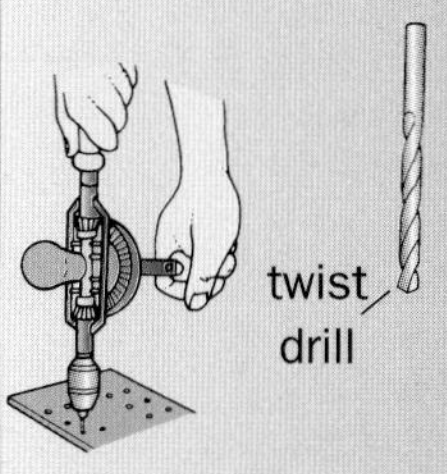 twist drill	W M	Difficult to use for large diameters	Cone cut	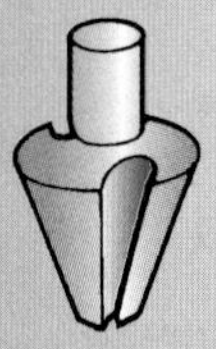	P M	Useful for enlarging holes in sheet plastic and metal
Brace/bits	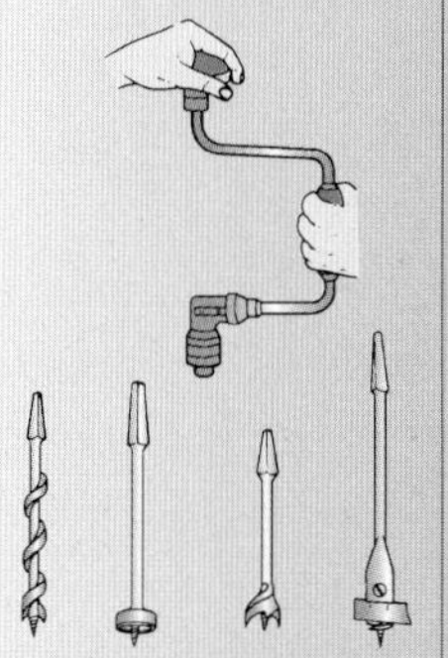 auger bit Forstener bit centre bit expansive bit	W	A wide range of bits are available for boring holes from 6 mm up to 50 mm diameter. Expansive (adjustable) bits can be used for larger holes	Bradawl	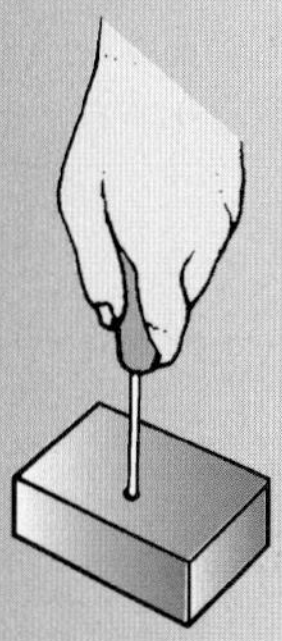	W P	For making small holes in timber and especially for 'starter' holes, prior to inserting woodscrews
Hole saw	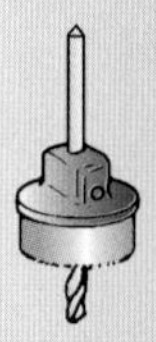	W M P	Use with a pillar drill at slow speed. Good for making wheels	Counter-sink drill		W M P	Used to countersink a hole

Fig 5.4 Hole-making tools   **Key:** W = wood, P = plastics, M = metals

Machine tools

AQA
EDEXCEL
OCR
WJEC
NICCEA

Depending on their availability and your school's **Health and Safety** policy, you may have access to the following machine tools. However, it is necessary for you to understand how they are used.

Goggles should always be worn when using machines, regardless of other safety devices provided.

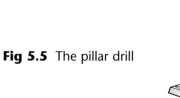

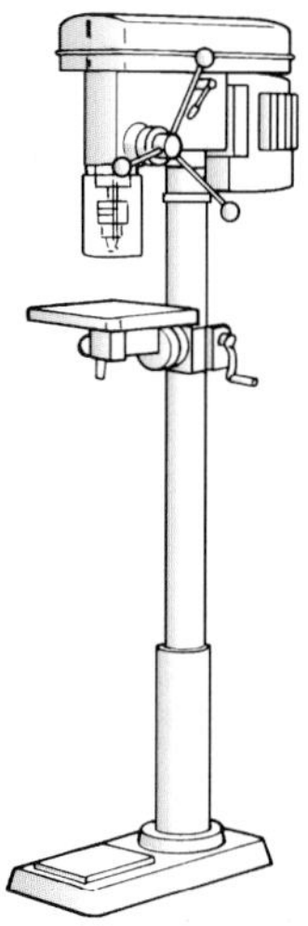

Fig 5.5 The pillar drill

The pillar drill

- This is used for accurately drilling holes to **set depths**.
- The speed can be **adjusted** to suit different materials and drill bit diameters.
- It is important that the work is **securely held**, especially thin sheet material as it has a tendency to spin.
- A particular hazard when drilling metal comes from the **swarf** (small, sharp coils of metal) which is frequently very hot and can burn the hands and face.

The vibrating (or jig/scroll) saw

- This is extremely useful for **intricate work** on **thin sheet** materials.
- To cut an **internal shape**, first drill a small hole in the work and then pass the blade through the hole before securing it to the saw.

It is essential to make sure that the pressure foot is correctly adjusted.

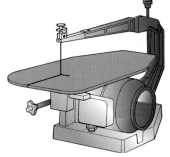

pressure foot/guard removed for clarity

Fig 5.6 The vibrating saw

The band saw

- This is used widely for cutting **curves and shapes** in timber, plastics and metal.
- A **special attachment** is available to aid the cutting of circles.

Finer blades are required for cutting metal.

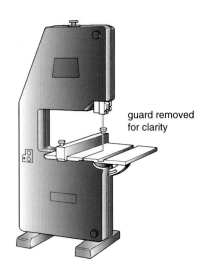

guard removed for clarity

Fig 5.7 The band saw

The wood-turning lathe

- This is used for turning **circular/cylindrical** objects in wood.
- The cutting tools (**chisels**) are held firmly by the operator and the diameter is progressively turned down.
- The work is either held on a small **faceplate** or **between centres**.

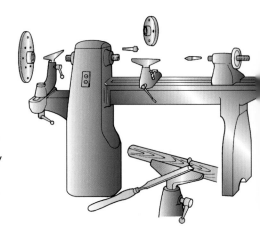

Fig 5.8 The wood-turning lathe

The centre lathe

- Sometimes referred to as a **metalwork lathe**, it is a versatile machine used for turning circular/cylindrical objects in **metal and plastics** (see Fig 5.9).
- The work is normally held in a three- or four-jaw **chuck** and supported if necessary.
- Once again, the **speed and tool selection** is critical in achieving a quality finish.

Great accuracy is achievable with practice.

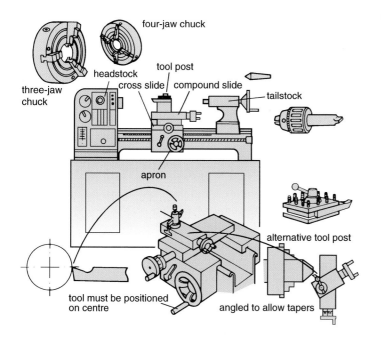

Fig 5.9 The centre lathe

The milling machine

● This is a very adaptable machine, designed to be used on **metals and plastics**.
● It is capable of producing flat surfaces, slots of various profiles, angled surfaces and accurately drilled holes.

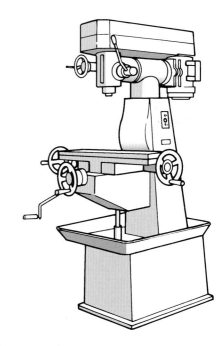

Fig 5.10 The vertical milling machine

KEY POINT

In industry the last three machines are likely to be computer controlled, capable of producing identical products in quantity.
CNC machines are often available in school and should be used where appropriate.

5.2 Materials

LEARNING SUMMARY

After studying this section you should be able to:
● *understand the difference between mechanical and physical properties*
● *select the appropriate material based on its properties*
● *consider the forms in which materials are available*

Properties of materials

AQA
EDEXCEL
OCR
WJEC
NICCEA

See Chapter 2, Coursework for more details on specifications.

All resistant materials have physical and mechanical properties that make them ideally suited for particular products. The properties required of a product should be detailed in the **design specification**, which will inform the choice of materials for a design.

Mechanical properties

Strength	The ability of a material to withstand an **applied force**
Hardness	A measure of how easily a material is **scratched or indented**. Very hard materials such as glass or cast iron are often **brittle** (they have low impact resistance)
Toughness	How well the material **absorbs impact** – the opposite to brittleness
Stiffness	The ability to **resist bending**
Ductility	The ability of a material to be **worked**. Ductile materials can be formed easily into shapes, e.g. pressing deep shapes in steel such as car body panels
Compressive strength	The ability to withstand **pushing** forces
Tensile strength	The ability to withstand **pulling** forces

Physical properties

Thermal conductivity	How well a material **conducts heat**, e.g. aluminium is a very good conductor of heat
Resistance to corrosion	How slowly the material **oxidises** – ferrous metals have a low resistance to corrosion
Electrical conductivity	How well the material conducts electricity – copper is a very good conductor. Materials that **do not conduct** electricity are called **insulators**, e.g. ceramics
Optical properties	How easily **light passes through** the material
Appearance	Considerations such as aesthetics, colour, brightness and texture
Joining properties	The ability of the material to be joined to itself or other materials – some woods, e.g. teak, are oily and therefore do not accept adhesives well

Material selection

Apart from the physical and mechanical properties of materials, you should also consider cost, availability (how easy it is to get hold of the material in the size you require), available tools and equipment, and your own expertise in the working processes necessary.

Ferrous metals

- The word **ferrous** comes from the Latin word for **iron**. As ferrous metals contain iron, they are almost all magnetic and, unless treated, rust very easily.

Material	Uses	Notes	Material	Uses	Notes
Cast iron	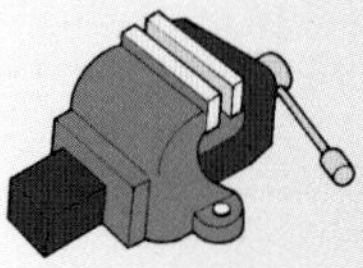	Hard skin. Strong under compression. Cannot be bent or forged	Stainless steel (alloy)		Hard and tough. Resists wear and corrosion. Quite difficult to cut or file
Mild steel		Tough, ductile and malleable. Easily joined but with poor resistance to corrosion. Cannot be hardened or tempered	High-speed steel (alloy)	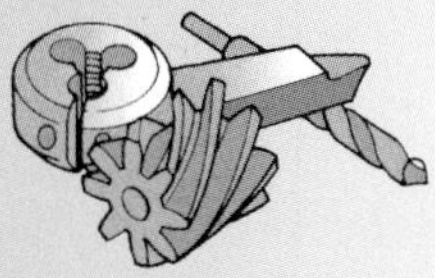	Very hard. Can be used as a cutting tool even when red-hot. Can only be shaped by grinding
High-carbon steel		Very hard but less ductile, tough and malleable. Difficult to cut. Can be hardened and tempered			

Fig 5.11 Ferrous metals

Non-ferrous metals

- These metals contain **no iron** and therefore **do not rust**.

The more common examples of non-ferrous metals are listed in Fig 5.12.

Material	Uses	Notes	Material	Uses	Notes
Aluminium		High strength/weight ratio. Difficult to join. Good conductor of heat and electricity. Corrosion resistant. Polishes well	Lead	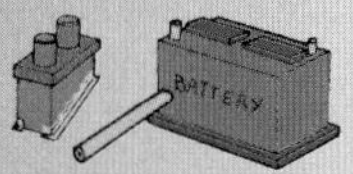	Very heavy, soft, malleable and ductile. Corrosion resistant. Low melting point. Difficult to work and expensive
Copper		Malleable and ductile. Good conductor of heat and electricity. Easily joined. Polishes well. Expensive	Tin (tin plate)		Soft and weak. Ductile and malleable. High corrosion resistance. Low melting point. Used to coat steel to produce 'tin plate'

Fig 5.12 Non-ferrous metals

Alloys

Metals that are made from a **mixture** of two or more metals or elements are called **alloys**.

- Alloys have different physical and mechanical properties from their constituent parts.

> **KEY POINT**
>
> Steel is one of the most common alloys available, being a mixture of iron and carbon. Iron is soft and ductile, while carbon is hard and brittle. By adjusting the amount of carbon used, steels with different hardness can be produced.

Material	Uses	Notes
Brass (alloy of copper and zinc)		Corrosion resistant. Harder than copper. Good conductor of heat and electricity. Polishes well. Cheaper than copper
Bronze (alloy of copper and tin)		Strong and tough. Corrosion resistant. Resistant to wear
Duralumin (alloy of aluminium, copper and manganese)		Nearly as strong as mild steel, but much lighter. Hardens with age. Machines well after annealing

Fig 5.13 Alloys

Timber

There are two different types of natural timber – **hardwoods** and **softwoods**. Natural timbers have a number of drawbacks in that they:

- suffer from changes in **moisture content** – they warp and twist if they are not properly seasoned (dried)
- are relatively expensive to buy
- do not come in **large sizes**, so several pieces will need to be joined to form a tabletop

Hardwoods

> **KEY POINT**
>
> Hardwoods come from trees that carry their seeds in fruit. They are broad-leaved and nearly always deciduous (they lose their leaves in the winter).

Although most hardwoods are hard, there are one or two exceptions, e.g. balsa wood.

- They are generally **slow growing** (typically 60–100 years to mature), with narrow growth rings that produce a **close-grained** timber that is usually strong and tough.
- However, because they are slow growing, they are much **more expensive**.

Material	Uses	Notes	Material	Uses	Notes
Mahogany		Easy to work. Fairly strong. Durable. Prone to warping	Teak		Hard, very strong and very durable. Very resistant to acids and alkalis. Contains grit, which blunts tools easily. Very expensive
Beech	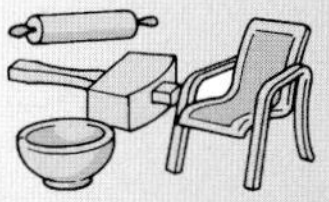	Close-grained, hard, tough and strong. Works and finishes well. Prone to warping	Jelutong	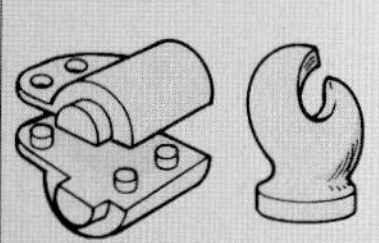	Pale cream in colour. Uniform grain. Shapes easily. Very few knots
Ash		Open-grained, tough and flexible. Good elastic qualities. Works well	Balsa	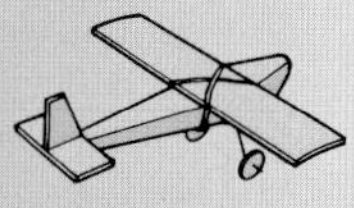	Very soft and light. Ideal for models. Quite expensive
Oak		Very strong, heavy and durable. Hard and tough. Open-grained. Contains tannic acid, which corrodes iron and steel			

Fig 5.14 Hardwoods

Softwoods

KEY POINT Softwoods come from trees that are cone bearing (coniferous). They have needles and are evergreen (they do not lose their leaves in winter).

Note that some softwoods, such as yew, are harder than many hardwoods.

- They are **fast growing** (20–30 years to mature), making them ideal for growing commercially. However, they are more **open-grained**, producing a weaker timber that can split easily.

- They are **much cheaper** to buy and can produce long, straight lengths with little waste on cutting. They are also used extensively in the paper industry.

Material	Uses	Notes	Material	Uses	Notes
Scots pine	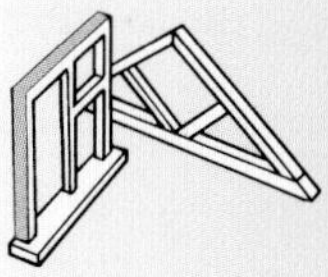	Straight-grained but knotty. Fairly strong, easy to work and cheap. Readily available	Western red cedar		Lightweight, knot-free, soft, straight silky grain. Durable against weather, insects and rotting. Easy to work but expensive
Spruce	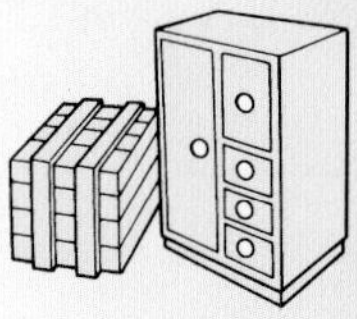	Fairly strong. Small hard knots. Resistant to splitting. Not durable	Parana pine	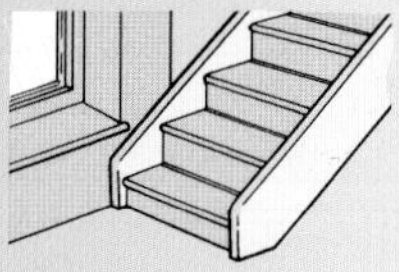	Hard, straight-grained. Almost knot-free. Strong and durable. Tends to warp. Expensive for softwood. Used for internal joinery
Douglas fir	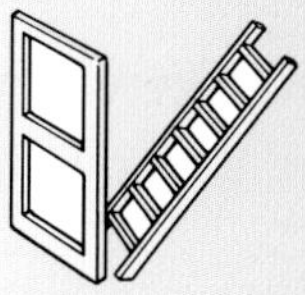	Dark red/brown. Fairly durable and quite dense			

Fig 5.15 Softwoods

Manufactured boards

- These can be produced in **large sizes** – up to 3 metres × 2 metres.
- They are **relatively cheap** and, because they do not have a grain running through them in the same way as natural timbers, they are **more stable**.

Material	Thickness available	Uses	Notes	Material	Thickness available	Uses	Notes
Plywood	3 mm 4 mm 6 mm 8 mm 10 mm 12 mm 15 mm 18 mm	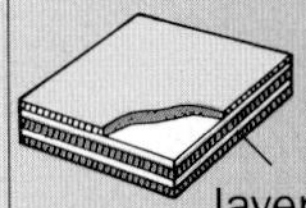 layers	Made from veneers of birch, alder, meranti or gaboon. Odd number of layers. Fairly cheap Much stronger than hardboard. Some forms of plywood resistant to moisture	Medium density fibre-board (MDF)	6 mm 12 mm 15 mm 18 mm	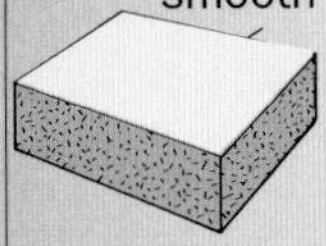 smooth	A sort of thicker, smoother, better quality hardboard. Has smooth faces and takes paint well. Best worked by machine
Chip-board	12 mm 15 mm 18 mm	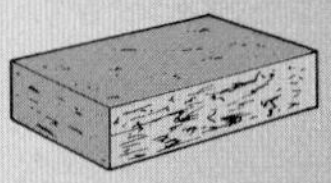	Chips of variety of timbers are bonded using synthetic glue. Available veneered with timber or plastic and used for cheaper, often 'flat pack', furniture	Block-board and lamin-board	12 mm 15 mm 18 mm	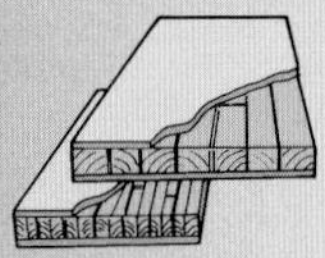	Cheaper to make, thickness-for-thickness, than plywood, although not the same uniform strength
Veneers	1 mm 2 mm 3 mm	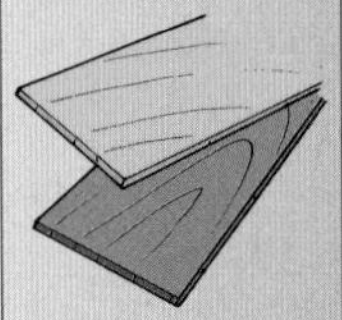	Thin layers of wood. Used for making plywood or laminating. A very economical use of timber since very little of the tree is wasted	Hard-board	3.2 mm 6 mm	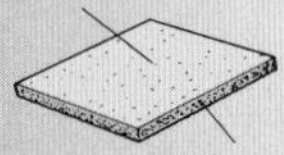 smooth rough	Cheap and fairly light. Used as a substitute for plywood. No grain. Equally strong in all directions. Standard hardboard absorbs moisture and must not be used outdoors. Usually smooth on one side only

Fig 5.16 Manufactured boards

Thermoplastics

- This is the most common type of plastic (**synthetic polymer**) used, known for its ability to adopt a new shape on heating, and then return to its original shape on reheating (**plastic memory**).
- Thermoplastics generally soften at a **low temperature** (as low as 100°C), making them inappropriate for situations requiring high temperatures.

Material	Uses	Notes	Material	Uses	Notes
Acrylic (PMMA)		Stiff, hard and uniform strength. Scratches easily. Clear; has good optical properties. Non-toxic. Good insulator, easily machined and polishes well	Acrylo-nitrile butadiene-styrene (ABS)	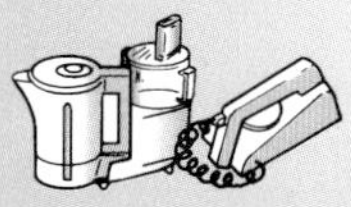	High impact strength. Tough and scratch resistant. Resistant to chemicals
			Polyvinyl chloride (PVC)	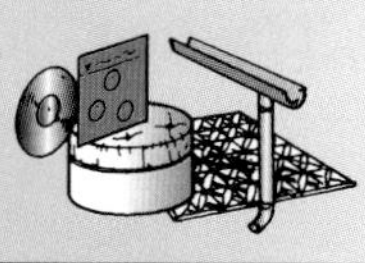	Chemical and weather resistant. Needs a stabiliser for outdoor use. Good electrical insulator
Rigid polystyrene (HDPS)		Light, hard, stiff, often transparent. Brittle with low impact strength. Water resistant. The toughened type can be coloured	Polyethyl-ene tereph-thalate (PET)		Used extensively for mineral water bottles. Clear and very tough
Expanded polystyrene (LDPS)		Buoyant, lightweight. A good sound and heat insulator	Polyethyl-ene (poly-thene, PE)		Tough, very popular. Quite cheap. Available in a wide range of colours. Fairly low melting point
Polyamide (nylon)		Usually creamy in colour. Hard, tough and resistant to wear. Low friction. Machines well, but very difficult to join			

Fig 5.17 Thermoplastics

> You are likely to have had workshop experience of using thermosets such as polyester resin (GRP or fibreglass) as well as epoxy resin adhesives (e.g. Araldite)

Thermosetting plastics (thermosets)

- When thermosetting plastics are being formed, a chain reaction occurs causing them to create strong links or **cross chains** in their structure, which means that once set they are **permanently formed** and cannot be reshaped on heating.

- Because they do not soften on heating, they can be used in situations where temperatures reach in **excess of 400°C**.

Material	Uses	Notes	Material	Uses	Notes
Polyester resin (GRP)	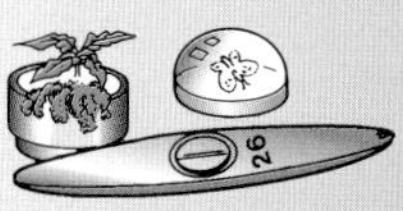	Stiff, hard and brittle. Used for casting and, when reinforced by glass fibres, produces GRP. Easy to colour. Excellent for outdoor uses	Epoxy resin	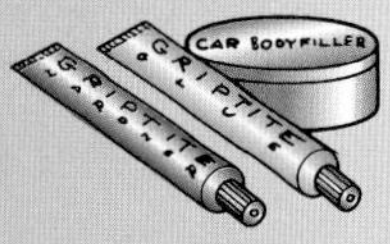	Very strong, especially when reinforced by glass or carbon fibres. Used as an adhesive for unlike materials e.g. metals to plastics
Urea formal-dehyde (UF)		Stiff, hard and brittle. Excellent electrical insulator. Used as an adhesive	Melamine formal-dehyde (MF)		Stiff, hard and strong. Scratch resistant. Low water absorption. Stain resistant. No odour. Available in a wide range of colours

Fig 5.18 Thermosets

Smart materials

> An example is nitinol, a shape memory alloy, which gives mechanical movement when a set temperature is reached.

Material technologists have developed materials capable of changing their properties, e.g. that change their length when subjected to electric current, or light-sensitive plastic. Many of these are **composites**, or have special **coatings**, e.g. temperature-sensitive coatings (thermocolour film). Other developments include conductive foams and plastics, and low temperature setting plastics (polymorph) that can be moulded using hot water.

Market forms

> **KEY POINT** When selecting appropriate materials, one consideration is the standard sizes or stock sizes that the materials are manufactured in.

Timber

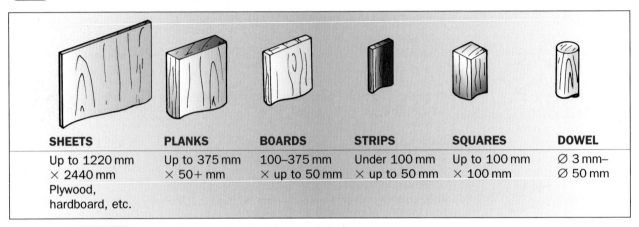

SHEETS	PLANKS	BOARDS	STRIPS	SQUARES	DOWEL
Up to 1220 mm × 2440 mm Plywood, hardboard, etc.	Up to 375 mm × 50+ mm	100–375 mm × up to 50 mm	Under 100 mm × up to 50 mm	Up to 100 mm × 100 mm	⌀ 3 mm– ⌀ 50 mm

Fig 5.19 Market forms of timber

> Timber is mostly sold by length, although large planks may be sold by the square or cubic metre.

Metal

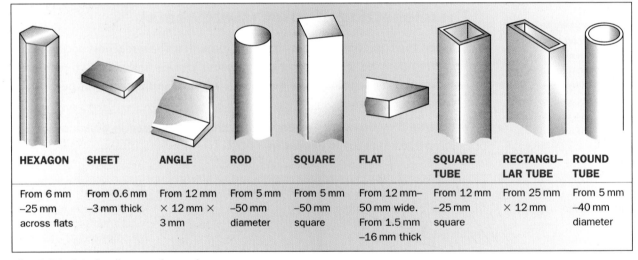

HEXAGON	SHEET	ANGLE	ROD	SQUARE	FLAT	SQUARE TUBE	RECTANGU–LAR TUBE	ROUND TUBE
From 6 mm –25 mm across flats	From 0.6 mm –3 mm thick	From 12 mm × 12 mm × 3 mm	From 5 mm –50 mm diameter	From 5 mm –50 mm square	From 12 mm– 50 mm wide. From 1.5 mm –16 mm thick	From 12 mm –25 mm square	From 25 mm × 12 mm	From 5 mm –40 mm diameter

Fig 5.20 Market forms of metal

Plastics

	POWDER	GRANULES	FOAM	FILM	SHEET	BLOCK	HEXAGONAL BAR	RODS	TUBES	RESINS AND PASTES
Polyethylene (PE)	✓	✓	✗	✓	✗	✗	✗	✗	✓	✗
Acrylic (PMMA)	✗	✓	✗	✗	✓	✓	✗	✓	✓	✓
Nylon	✓	✓	✗	✗	✗	✓	✓	✓	✓	✗
PVC	✓	✓	✗	✗	✓	✗	✗	✗	✓	✗
Polystyrene (PS)	✓	✓	✓	✗	✓	✓	✗	✗	✗	✗
Polyester	✗	✗	✓	✗	✗	✗	✗	✗	✗	✓
Epoxy	✗	✗	✗	✗	✗	✗	✗	✗	✗	✓

Fig 5.21 Market forms of plastics

5.3 Construction

LEARNING SUMMARY

After studying this section you should be able to:

- **select the most appropriate form of adhesive**
- **distinguish between the soldering and welding of metals**
- **select the most appropriate joint in wood**
- **use knock-down and mechanical fittings**

Adhesives

AQA
EDEXCEL
OCR
WJEC
NICCEA

The selection of the most appropriate adhesive depends on a combination of the following requirements.

- The **materials** to be joined
- The **working strength** required
- The **environment** where it will be used

Table 5.1 Adhesives

Adhesive	Uses		Hints on use	
Polyvinyl acetate (PVA) Evostik resin W	Used mainly for timber and paper products.	Long setting time (several hours)	Not always waterproof. Wipe off excess before it dries.	Sustained pressure is needed
Synthetic resin Cascamite Aerolite 306	Stronger than PVA and also waterproof.		Chemically active. Needs mixing with water. Will fill small gaps in a joint.	
Acrylic cement Tensol 12	Used only for acrylic. It does not work with other plastics or other materials.		Ensure good ventilation. Replace cap when not in use.	Usually only hand pressure is needed
Epoxy resin Araldite	Expensive but versatile. Will bond almost any clean material.	Short setting time (less than one hour)	Resin and hardener need to be mixed. It hardens quite quickly but does not reach full strength for two to three days.	
Contact adhesive Evostik Contact Thixofix (Dunlop)	Used mainly for glueing sheet material, such as melamine to work surfaces.		Apply thin layers to each side. Allow to dry. Adhesion occurs on contact. The vapours are harmful and ventilation is essential.	
Latex adhesive Copydex	Suitable for fabrics, paper and upholstery.		Non-toxic and safer for young children to use.	
Polystyrene cement Airfix cement	Used only on rigid polystyrene (expanded polystyrene will melt).		Ensure good ventilation.	
Rubber solution Bostik	Used only for rubber, especially in bicycle puncture repairs.		Read the manufacturer's instructions carefully.	
Glue gun	Used for rapid glueing of small pieces.		Take care, as the glue is used hot and can burn the skin badly.	

Soldering and welding metals

KEY POINT

Although permanent joints in metal can be made using adhesives, the working strength required usually necessitates a soldered or welded joint. Both processes involve heat and the addition of a second metal, which melts to form the joint.

Soldering

> Care must be taken to protect the electronic components from overheating.

> All soldering requires the use of a flux to keep the joint clean and to help the solder flow.

There are two main types of soldering.

- **Soft soldering** is used for attaching (soldering) electric components to a circuit board and joining thin plate. It uses a solder (the name given to the filler material) made from **lead and tin** (40%/60%) which has a low melting point (**180°C–200°C**).

- **Hard soldering** (brazing and silver soldering) produces much stronger joints but requires more heat (**625°C–875°C**). Brazing uses a spelter (**65% copper/35% zinc**), while silver soldering uses **silver** (50%–81%) alloyed with **copper and zinc**.

soft soldering electronics

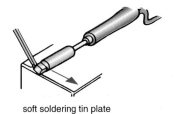
soft soldering tin plate

hard soldering

Fig 5.22 Soldering techniques

Welding

- This process involves much greater heat than soldering, often by the use of an **oxygen/acetylene torch**. A **filler rod** of the same material fills the gap between the two surfaces.

- An **electric arc welder** is often used to weld steel. A very **high current** is used to create an arc between the rod and the metal to be joined (which is connected to earth). The heat from the arc melts the metals and fuses the joint together.

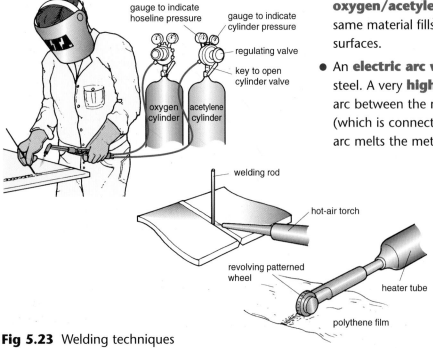

gauge to indicate hoseline pressure
gauge to indicate cylinder pressure
regulating valve
key to open cylinder valve
oxygen cylinder
acetylene cylinder
welding rod
hot-air torch
revolving patterned wheel
heater tube
polythene film

> Plastic can also be welded using heat (around 400°C). Heat welding is used for sealing polythene packaging.

Fig 5.23 Welding techniques

Joints in wood

AQA
EDEXCEL
OCR
WJEC
NICCEA

There are many ways of permanently joining wood. The choice of method depends on the function of the product being constructed. There are three main reasons for using construction joints.

- **Strength** – many joints 'lock' the two pieces of wood together when accurately produced, e.g. dovetail joint. This is particularly important when forces such as compression, shear and tension act upon the joint.
- **Appearance** – many construction joints look good and contribute to the aesthetic qualities of the design.
- **Quality** – because of improved strength and appearance, they add quality to a product.

KEY POINT

The strength of a glued joint depends largely on the size of the glueing area – cutting a joint increases the contact area of the wood.

Name	Joint	Advantages/disadvantages	Name	Joint	Advantages/disadvantages
Butt	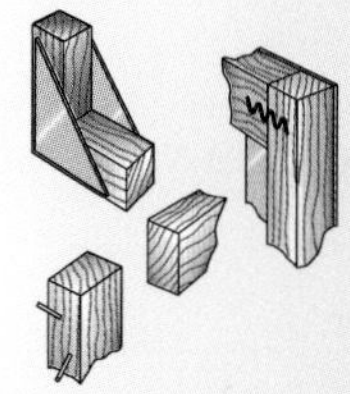	Simple, very little preparation needed. Low strength	Dovetail joint	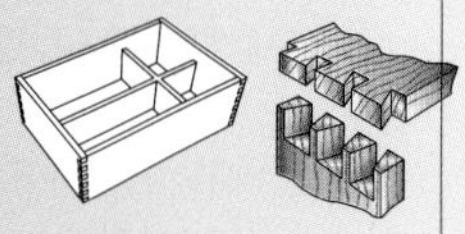	Strongest corner joints. Have a large glueing area, lock the two pieces together and can only be pulled apart in one direction. Although difficult to make they look very attractive and are often found on quality furniture
Mitre	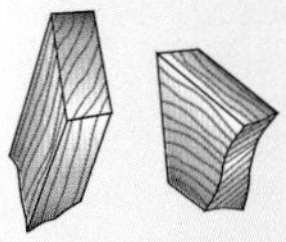	Attractive, often used as a decorative corner joint, hides the end grain. Low strength. Harder to make than the butt as it involves cutting the edges accurately at 45°	Housing joints	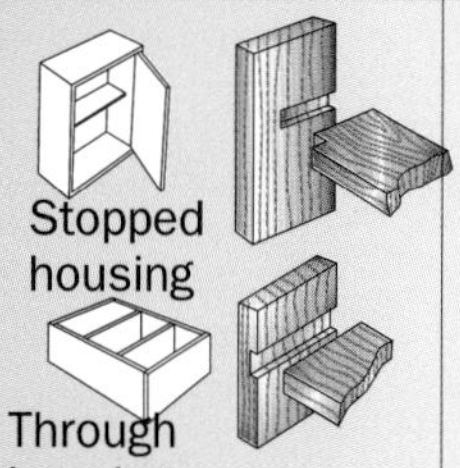 Stopped housing / Through housing	Used for shelves and partitions. If the front of the joint is not to be seen, a stopped housing is used. Stronger than butt joints
Dowelled joints	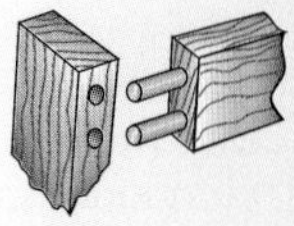	Neat and strong. The holes for the dowels must be lined up extremely accurately. This can be done using a dowelling jig	Comb/finger joints	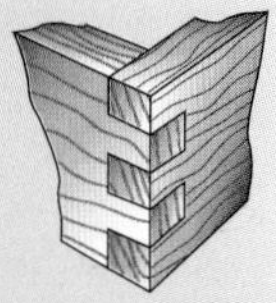	Much easier to cut than dovetails. They are strong as a result of the large glueing area. They are also attractive
Corner halving	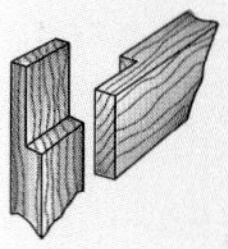	Relatively strong and quite easy to make. Usually needs strengthening with either screws, dowel or a gusset	Mortise and tenon	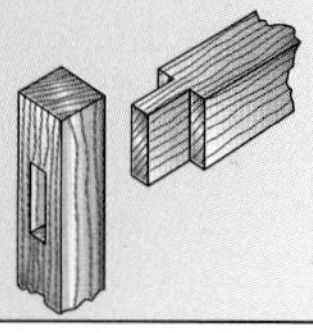	The strongest and most important joint used in frame construction. Used for strength in doors and furniture

Fig 5.24 Types of wood joint

Fittings and fixings

AQA
EDEXCEL
OCR
WJEC
NICCEA

Knock-down fittings

- These are used when the joint is temporary, semi-permanent or the material is inappropriate for permanent joining, e.g. plywood and chipboard split very easily.

Knock-down fittings are widely used in 'flat-packed' furniture.

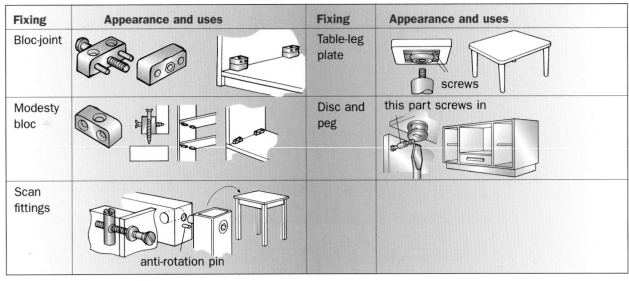

Fixing	Appearance and uses		Fixing	Appearance and uses	
Bloc-joint			Table-leg plate	screws	
Modesty bloc			Disc and peg	this part screws in	
Scan fittings	anti-rotation pin				

Fig 5.25 Knock-down fittings

Mechanical fittings

- These are normally used for semi-permanent or temporary joining. There is a wide variety, and each has a particular use.

Fixing	Appearance	Uses	Fixing	Appearance	Uses
Round wire nail	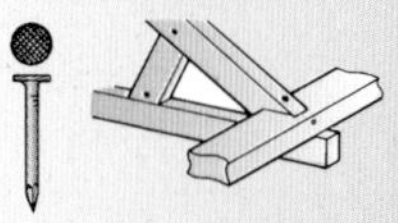	General purpose nail. Do not use near end of timber or timber will split	Nut and bolt	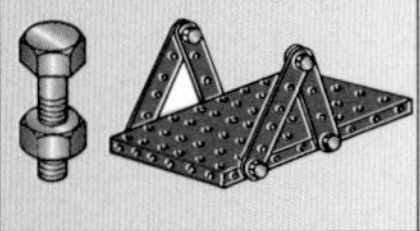	General purpose. Tightened with spanners. Can be very strong
Oval wire nail	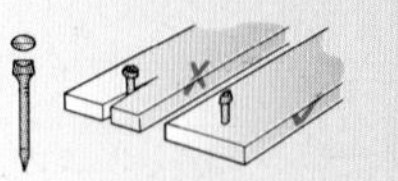	Similar use to round wire nail, but less likely to split wood if used correctly. More likely to bend	Self-tapping screw	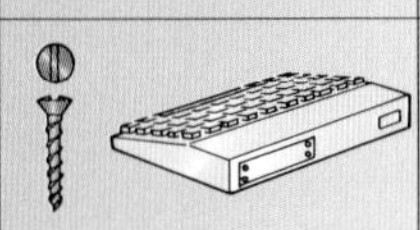	Used to cut a thread in sheet metal and soft plastics. A small pilot hole is needed
Hardboard nail	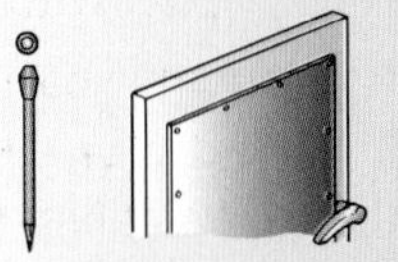	Designed to fix panels of hardboard. The head is shaped so that it sinks into the surface	Rivets	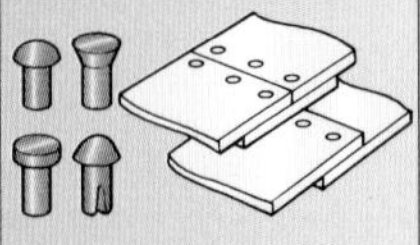	Available in a range of materials, but frequently with countersunk or snap heads (as shown). Used to fix metal components together
Masonry nail		Useful for quick (and sometimes crude) fixing to brick, concrete and block walls	Pop rivet		Used in sheet metal and fixed with a special tool. Used when it is impossible to reach both sides of the metal
Counter-sunk woodscrew		Usually with a slotted or cross-point head. Used to fix two pieces together. Make sure the piece nearest to the head has clearance-sized hole drilled	Butt hinge	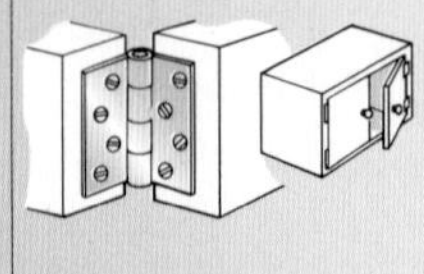	A general purpose hinge used to allow movement on doors, for example
Round-head wood-screw	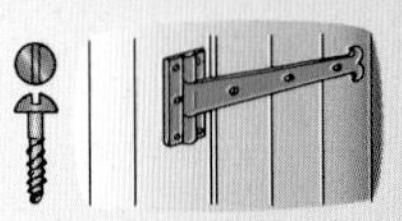	For holding panels of thin material or where it is not possible to make a countersink	Kitchen cabinet hinge		Used for modern kitchen fittings. Hinge is invisible from outside
Clout nail		For holding roof felt. Galvanized for outdoor use			

Fig 5.26 Mechanical fittings

5.4 Surface finishes

After studying this section you should be able to:

● *select the most appropriate finish for a product*

The surface finish you apply to resistant materials has a substantial effect upon the **quality** of the product. Surface finishes or treatments have both a functional and an aesthetic aspect.

> **The required finish for a product will be identified in the specification.**

● **Function** – to protect the product. This could be to prevent oxidisation, scratching or tarnishing.
● **Aesthetics** – colour can create an image or style, it can imbue a sense of quality, make a product look heavy, light or even stand out.

Before choosing an appropriate surface finish, there are five things to consider.

● The type of material
● The function of the finish
● How it is going to be applied
● The skill of the person applying the finish
● The cost

Finish	Uses	Notes	Finish	Uses	Notes
Paint and primer		Take care to prepare surface. Remove rust or old paint before applying primer	Plastics coating	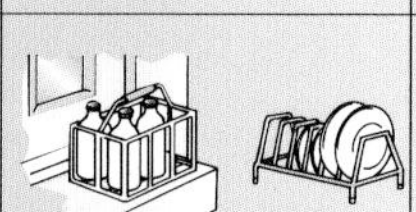	Commonly PVC and polythene are used to 'dip' steel. This is done in a fluidising tank
Cellulose paint	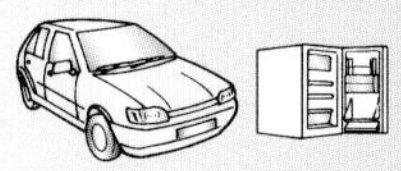	Usually sprayed with specialist equipment. Care must be taken to avoid breathing fumes	Enamelling		Enamels are applied to copper surfaces, to which they 'fuse' after heating
Lacquer		Material such as brass or copper, which will discolour after a time, can be protected by a coating of lacquer	Varnish	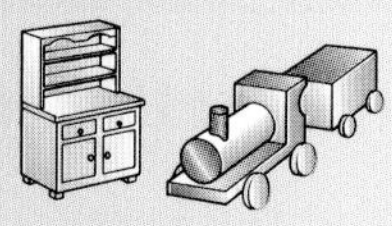	Prepare surface thoroughly. Apply two thin coats. Rub down between coats. Clean brushes with white spirit
Tin plating		Tin plating protects the surface of steel from corrosion and protects the contents, e.g. food, from contamination by the steel	Primer, undercoat and paint		Top coat will only stick properly to an undercoat. The primer is designed to fill the grain of the timber
Chrome plating		Chrome plating gives a hard, shiny attractive finish. It can be applied to steel or brass	Cellulose spray paint		Aerosols containing CFCs should be avoided. Cellulose vapour is harmful. You must ensure good ventilation and wear a face mask

Fig 5.27 Finishes

Finish	Uses	Notes	Finish	Uses	Notes
French polish	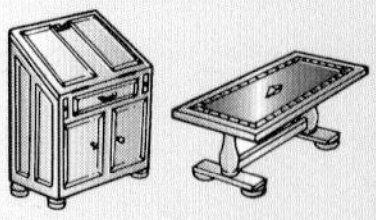	A difficult process best left to the experts. The finished surface is of a high gloss but is easily damaged by heat	Oil	Teak oil  Olive oil Linseed oil	Oils provide good resistance to water. They are quick and easy to apply

Fig 5.27 Finishes (continued)

5.5 Manufacturing processes

LEARNING SUMMARY

After studying this section you should be able to:

● *understand the processes used to mould plastics*
● *understand the casting process for aluminium*

> **KEY POINT** It is important that you make yourself aware of some of the most widely used industrial manufacturing processes in the main resistant materials.

When deciding upon the most appropriate manufacturing process, designers need to consider:

● how many products are going to be produced and at what rate – the scale of production
● the form of the product – shape, intricacy, complexity
● the material to be used

Plastic moulding

AQA
EDEXCEL
OCR
WJEC
NICCEA

Examples include casings for electrical equipment and watches.

A complex mould could easily cost £100,000.

Injection moulding

● This process involves injecting molten **thermoplastic** into a **mould** under great pressure. The moulds can be very intricate and are often made in **several pieces** to allow the item to be removed.

● Injection moulding is widely used for making products in **volume** very cheaply. This is because the initial tooling costs to make the mould are too high for one-off or small batch production.

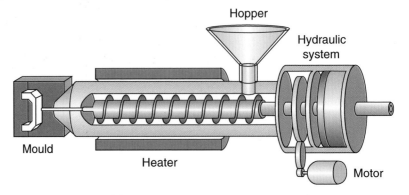

Fig 5.28 Injection moulding

Extrusion

- This process involves forcing a molten **thermoplastic** through a **die** and rapidly cooling the resultant shape as it emerges.

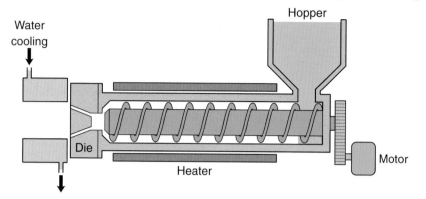

- This is commonly used for making products such as pipes, tubes, guttering and UPVC window frames.

Fig 5.29 Extrusion moulding

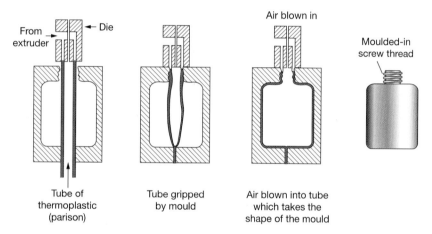

Fig 5.30 Extrusion blow moulding

Blow moulding

- The process is similar to glass blowing and is used for the industrial production of **bottles and containers**.

- A tube of hot (therefore flexible) **extruded thermoplastic** is gripped at both ends by the mould and **air** is blown into the tube, which then expands to take on the shape of the mould – including any relief details such as threads and surface decoration.

Vacuum forming

- A flat sheet of **thermoplastic** is clamped above a mould and heated until it becomes very flexible. The air is then pumped out from below the sheet allowing it to be **drawn down** over the mould (mainly due to atmospheric pressure), thus taking up its shape.

- Vacuum forming is used commercially for producing large components such as car dashboards, baths from acrylic, and packaging for goods.

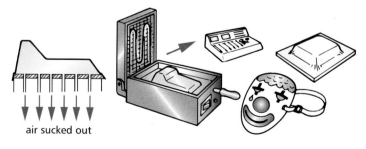

air sucked out

Fig 5.31 Vacuum forming

> **KEY POINT**
> Because all of the above processes use thermoplastics, there is little waste produced. All scrap can be chopped up and used again, which is not possible with thermosetting plastics.

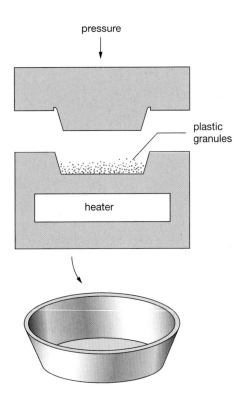

Fig 5.32 Compression moulding

Compression moulding

- This process involves placing a measured 'plug' of mixed **thermoset** between the two halves of a split mould. The mould closes under heat and pressure, and the moulding is **'cured'** for approximately two minutes, during which time the cross links are formed.
- Compression moulding is widely used to produce components such as fan housings, electrical plugs and sockets.

GRP (fibreglass) moulding

- This is another widely used process involving the use of **thermosets** in the production of **glass reinforced polyester**.
- This composite has high tensile and compressive strength, is light, hard wearing and has excellent resistance to corrosion.

> Used for small boats, exterior casings and some car body panels.

Casting in aluminium

> This process can be carried out in school workshops under strictly supervised and controlled conditions.

Sand casting

- This process involves making a wooden mould (flat or split), called a **pattern**, which is the same shape as the object to be cast.
- The pattern is packed in special sand within two steel boxes **(cope and drag)**. These are then split for the pattern to be removed, leaving a perfect impression in the sand.
- Molten aluminium is then poured in through preformed **sprue** holes known as **runners and risers** which are formed by removeable **sprue pins**.

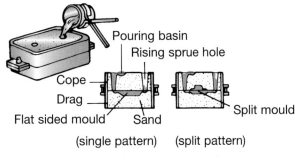

Fig 5.33 Sand casting in aluminium

Advantages	Disadvantages
• Low initial outlay	• A new sand mould needs to be made every time
• Ideally suited to large products which do not require high levels of accuracy	• Relatively slow • Casting requires finishing
• Pattern can be reused	

Die casting

This is a similar process to injection moulding of plastics.

- This process, also known as **gravity die casting**, uses moulds made from **high carbon steel** which can be used repeatedly.
- The high initial cost is recouped by **higher production rates**, which can be increased further by using **pressure die casting**.

5.6 Quality

LEARNING SUMMARY

After studying this section you should be able to:

- *understand how quality control and quality assurance are achieved*

Quality control

AQA
EDEXCEL
OCR
WJEC
NICCEA

Quality control involves **testing**, usually by **sampling** at a predetermined rate, to ensure that the specified standards are being maintained during manufacture.

- When the working drawing is produced it will contain all the important dimensions, including a **tolerance** – the amount by which a component can vary in size from the **absolute design size**.
- **Quality controllers** will often use specially designed gauges to speed up the sampling, with the two limits of the tolerance accurately set. These are known as the **'go'** and the **'no go'** dimensions.

KEY POINT

Tolerance is usually stated as + or −, e.g. **100 mm +/− 1 mm**. This would mean that the component is acceptable if it lies between **99 mm and 101 mm**.

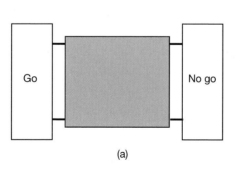

(a)

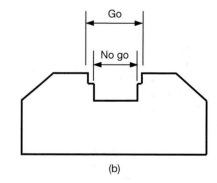

(b)

Fig 5.34 Gauges (a) Plug gauge for checking the diameter of a hole (b) Length gauge

Quality assurance

AQA
EDEXCEL
OCR
WJEC
NICCEA

See Chapter 1, p. 10 for further information on industrial practices.

One important aspect of quality assurance is ensuring **accuracy and consistency**. The most common way of ensuring this consistency is to use **jigs and fixtures**.

Jigs

If, for example, four holes are to be drilled accurately and consistently in 1000 pieces of steel as shown in Fig. 5.35, a jig would need to be designed that would:

- **clamp** the work in the same position for each piece
- **guide** the drill bit

The steel plate would be clamped within a jig as shown in Fig 5.36. The drill bit is guided by the **hardened steel bushes**, which are used to prevent wear and so they can be replaced, if necessary, to extend the life of the jig.

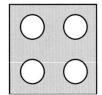

Fig 5.35

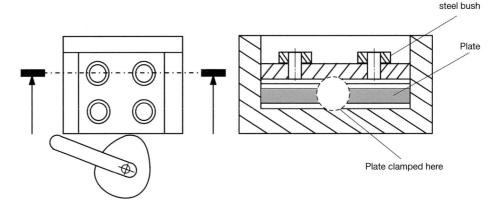

Fig 5.36 Example of a jig in use

Fixtures

A fixture differs from a jig in that it is **fixed to the bed** of the machine tool, whereas a jig is **moveable** in order to line up with the tooling.

5.7 *Mechanisms*

After studying this section you should be able to:

- *distinguish between the different types of mechanism*
- *perform simple calculations*

KEY POINT

A mechanism is a system that transforms one kind of motion to another.

There are five basic types of mechanism.

Levers	Enable a small amount of force to exert a larger force at a particular point
Linkages	Connect different systems together and have the important function of changing the direction and size of the force. They can also produce parallel movement
Rotary systems	Include gears, pulleys and belts, chains and sprockets which transmit force, and can change the speed and direction of movement
Cams and cranks	Convert rotary motion to linear reciprocating movement
Screws	Allow a rotary motion to exert a linear (straight-line) force, e.g. a G-cramp or vice

Levers

AQA
EDEXCEL
OCR
WJEC
NICCEA

These are simple mechanisms which create an 'advantage' for the user. There are three key parts to all levers.

- **Effort** – the force exerted by the user
- **Load** – the force exerted by the object being acted upon
- **Fulcrum (or pivot)** – the point at which the lever acts

There are three types of lever.

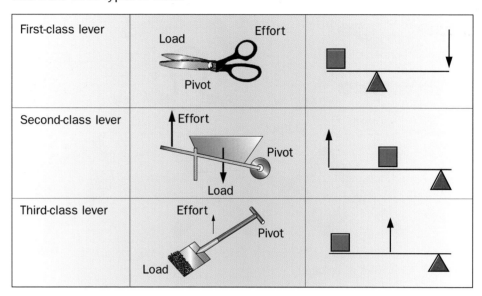

First-class lever	
Second-class lever	
Third-class lever	

Mechanical advantage

- The mechanical advantage (MA) is the ratio of the load (output force) divided by the effort (input force).

$$MA = \frac{Load}{Effort}$$

Velocity ratio

Mechanical advantage, velocity ratio and efficiency can be used on any mechanism. They are not restricted to levers.

- The velocity ratio (VR) is the ratio of the distance moved by the effort divided by the distance moved by the load.

$$VR = \frac{\textbf{Distance moved by the effort}}{\textbf{Distance moved by the load}}$$

Efficiency

- The efficiency of a system can be calculated using mechanical advantage and velocity ratio.

$$\textbf{Efficiency} = \frac{\textbf{MA}}{\textbf{VR}} \times \textbf{100\%}$$

KEY POINT Although the efficiency should theoretically be 100%, friction and other losses will always reduce efficiency.

Linkages

AQA
EDEXCEL
OCR
WJEC
NICCEA

These are very useful mechanisms as they allow forces and motion to be transmitted to where they are needed, often at some distance from where they are generated.

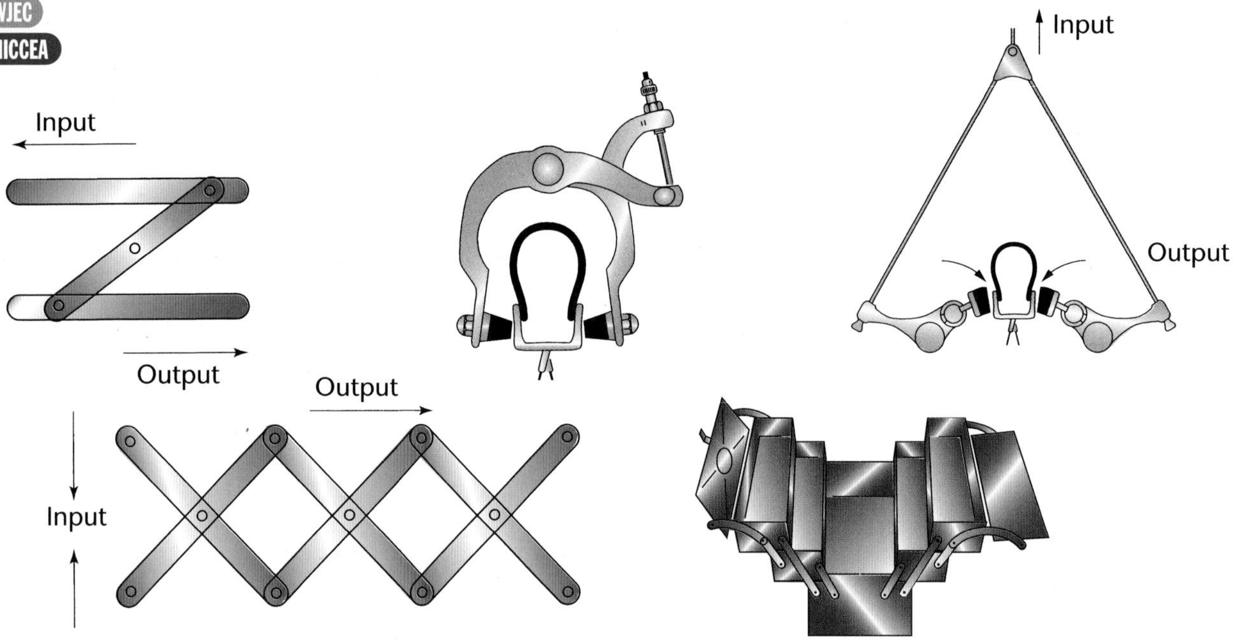

Input

Output

Output

Input

Input

Output

Fig 5.37 Examples of linkages

Rotary systems

AQA
EDEXCEL
OCR
WJEC
NICCEA

There are three main types of rotary mechanism used to transmit forces.

● Gears
● Belts and pulleys
● Chains and sprockets

Gears

These are used in situations where force needs to be transmitted **without slip**.

● Gears are toothed wheels that are fixed to the driver shaft and the driven shaft – which are usually **parallel** to each other.

● Two **meshed** gears will turn in **opposite** directions. By using an **'idler' gear** on a separate parallel rotating shaft, the driver and driven gears can be made to turn in the **same** direction, without affecting the **gear ratio** – see Fig 5.38.

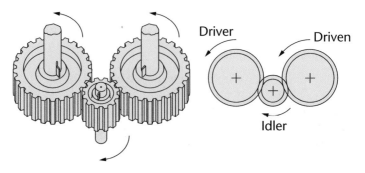

Driver

Driven

Idler

Fig 5.38 Pictorial and symbolic representation of the use of an idler gear in a simple gear train

Gear ratio

This is the method used to calculate the **change in speed of rotation** of the gears.

Velocity ratio or Gear ratio $=$ **Number of teeth on the driven gear** / **Number of teeth on the driver gear**

- If the gear ratio is **greater than 1**, there will be a **velocity decrease**.
- If the gear ratio is **less than 1**, there will be a **velocity increase**.

Simple gear trains

A simple gear train has only **one gear** on each parallel shaft. Regardless of the number of meshed gears, the gear ratio for each shaft will still be the ratio of the number of teeth on the gear attached to it (driven), divided by the number of teeth on the gear attached to the first shaft (driver). (See Fig 5.38)

Compound gear trains

A compound gear train will have at least **two gears** attached to one of the shafts (see Fig 5.39), and can achieve large changes in speed of rotation.

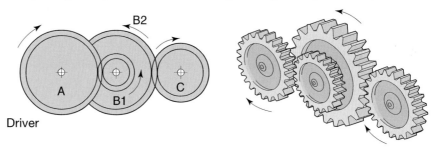

Fig 5.39 A compound gear train

In this case, the gear ratio is calculated by **multiplying the individual gear ratios** for each shaft together, e.g.

$$\text{Gear ratio} = \frac{\text{Driven B1}}{\text{Driver A}} \times \frac{\text{Driven C}}{\text{Driver B2}}$$

> **KEY POINT**
>
> As the speed decreases (by **stepping down** the gears), the **torque** of the driven gear increases.

Bevel and wormwheel gears

By using these gear arrangements, the driver and driven shafts need not be parallel to each other.

The worm gear cannot be driven by the wormwheel.

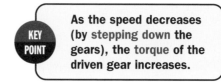

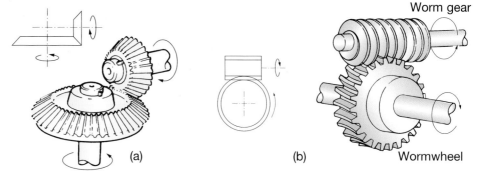

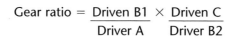

Fig 5.40 (a) Bevel gears (b) Worm gear and wormwheel

Rack and pinion gears

Using a special gear mechanism called a rack and pinion, gears can be used to convert **rotary motion to linear motion**.

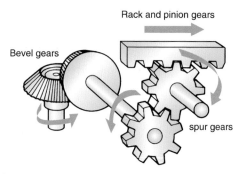

Fig 5.41 Rack and pinion

Belts and pulleys

Pulleys use **vee belts** to transmit force and, unlike gears, the driver and driven pulley rotate in the same direction – unless the belt is **'crossed'** (See Fig. 5.42).

- Stepped cone pulleys are used on machines like a pillar drill to achieve **different speeds** using just **one belt** (See Fig. 5.43).
- To **avoid slipping**, and where accuracy of rotation is required, a **toothed belt** can be used (See Fig. 5.44).

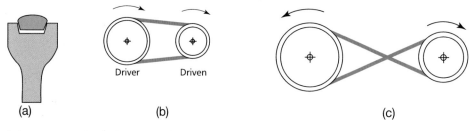

Fig 5.42 (a) Cross-section of a vee belt and pulley **(b)** Pulleys set up normally **(c)** Pulleys set up with a crossed belt

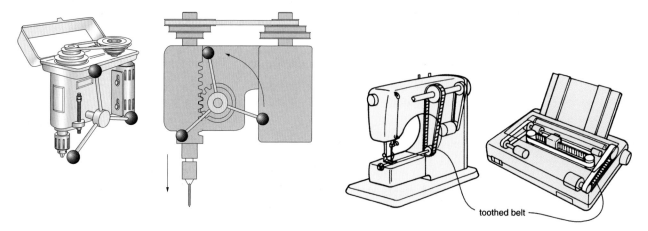

Fig 5.43 Stepped cone pulleys on a pillar drill   **Fig 5.44** Toothed belt applications

Advantages of belts and pulleys	• Belts are much quieter than metal gears
	• Belts do not need to be lubricated
	• Belts allow some fitting tolerance
	• Flat and vee belts allow some slippage as a matter of safety
	• Easier to replace
Disadvantages	• They can slip
	• They are more prone to wear
	• Can only be used in clean/dry conditions

Chains and sprockets

A sprocket is a toothed wheel driven by a chain (a series of metal links).

- Bicycles and motorbikes use sprockets and chains because of their greater **strength** and the fact that they do not slip.

Fig 5.45 Use of sprocket and chain

Advantages of chains and sprockets over belts and pulleys	• Larger forces can be transmitted • They do not slip • The links can be taken apart and removed for maintenance

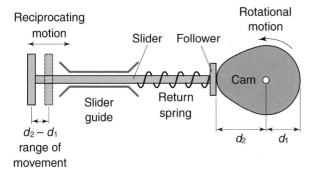

Fig 5.46 The basic cam

Cams

Cams are widely used mechanisms, which convert **rotary motion** into **linear motion**.

- As the cam rotates, the follower moves up and down (**reciprocating motion**) in a manner dependant on the **shape** of the cam.
- The movement of the other end of the follower can then be used to **control another mechanism**, e.g. a valve.

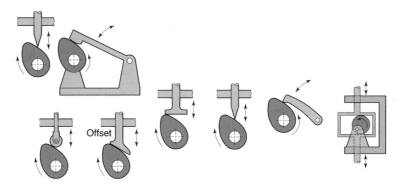

Fig 5.47 Some examples of cam and follower shapes

Crank and slider

A crank and slider is another method for converting rotary motion into linear (reciprocating) motion.

- The distance travelled by the slider is dependent on the **radius of the crank**.
- The crank and slider mechanism is used in engines (the crankshaft and piston) and on machines such as a power hacksaw.

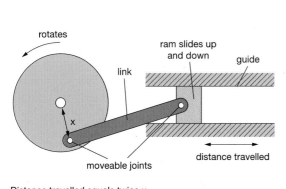

Distance travelled equals twice x

Fig 5.48 The crank and slider mechanism

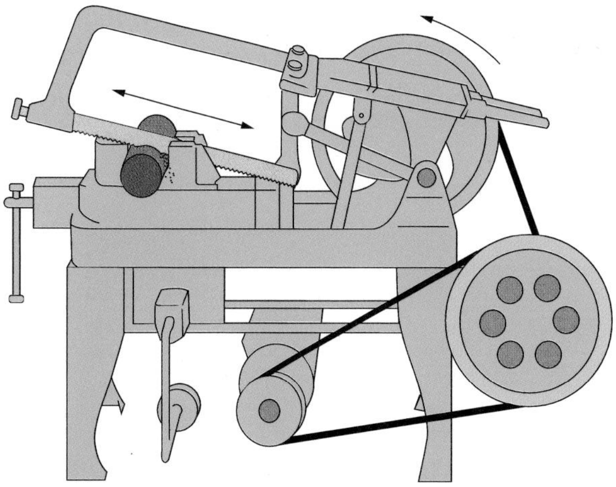

Fig 5.49 The power hacksaw

Systems and control technology – electronic products

The following topics are covered in this chapter:

- **Tools and equipment**
- **Basic electrical concepts**
- **Basic electronic components**
- **Timing circuits**
- **Amplifiers**
- **Alternating current**
- **Logic and counting**
- **Pneumatics**

The different Examination Boards have various approaches to this area of subject knowledge – see the summary on pages 4–5. The most common is **Electronic Products**, which forms the basis of this chapter. However, others such as Systems and Control Technology draw on **Pneumatics**, which is included in this chapter, and **Mechanisms**, which can be found in Chapter 5.

6.1 Tools and equipment

LEARNING SUMMARY

After studying this section you should be able to select, name and use appropriately:

- **equipment and tools used to construct electronic circuits**
- **tools and processes for working with resistant materials**

Designing and manufacturing electronic circuits

AQA
EDEXCEL
OCR
WJEC
NICCEA

Whilst it is likely that you will only have direct experience of working in a domestic/school environment, you should also understand the applications on an **industrial** scale.

You will be expected to acquire the skills necessary to use effectively and safely the following tools, equipment and processes.

- Computer-aided design packages for circuit design
- PCB manufacture, including etch tank
- Soldering iron
- Test equipment
- Resistant material processes

See Chapter 5, Resistant materials technology for more information on resistant materials.

Although the main focus of your coursework will be the electronic/control systems, you must ensure that, in your folder, you take full account of the design aspects of your **product**.

This will include such aspects as:

- aesthetics
- ergonomics
- safety

6.2 *Basic electrical concepts*

After studying this section you should be able to:

● **understand the basic electrical units**
● **recognise the numerical units used**
● **use Ohm's law**

Electrical units

AQA
EDEXCEL
OCR
WJEC
NICCEA

There are five basic electrical units.

The volt (Symbol – V)

● This is the unit of **electrical pressure**, meaning the **difference in potential** between the **positive and negative** terminals of the power source.
● The higher the voltage, the greater is the force acting to cause current to flow in a given circuit.

The amp (Symbol – A)

> In most circuits, the amp is far too large a unit and the milliamp – one thousandth of an amp – is used.

● This is the unit of **current**, meaning the **rate** at which electricity flows around a circuit.
● The higher the current, the 'faster' the current flows through a circuit.

The ohm (Symbol – Ω)

> Copper is the most commonly used conductor because it has a very low resistance.

● This is the unit of **resistance**, since all materials (except superconductors) oppose the flow of electricity.
● Materials that have a very **high resistance are called insulators**, whilst materials that have a very **low resistance are called conductors**.

The watt (Symbol – W)

> The watt is usually too large a unit to be used in microelectronics and so the milliwatt – one thousandth of a watt – is used.

● This is the unit of **power**, which in electronics is the voltage x current.
● **Power = I (amps) $\times$ Volts**.

The farad (Symbol – F)

> The farad is an impractical unit, so the μF is used.

● This is the unit of **capacitance**, which is the amount of electrical charge that can be stored in a **capacitor**.

Numerical units and symbols

AQA
EDEXCEL
OCR
WJEC
NICCEA

The following symbols and prefixes are used:

Three other symbols are commonly used.
Current = I
Resistance = R
Power = P

Symbol	Prefix	Multiplier	
G	giga	one thousand million	10^9
M	mega	one million	10^6
k	kilo	one thousand	10^3
m	milli	one thousandth	10^{-3}
μ	micro	one millionth	10^{-6}
n	nano	one thousand millionth	10^{-9}

Ohm's law

AQA
EDEXCEL
OCR
WJEC
NICCEA

Ohm's law is a very important principle to learn as it defines the relationship between the voltage (V), the current (I), and the resistance (R).

- A simple way of remembering this formula in order to find the missing unit easily is to use the '**magic triangle**'.

Therefore: $V = I \times R$ and $I = \dfrac{V}{R}$ and $R = \dfrac{V}{I}$

KEY POINT Voltage = I (current) × Resistance

6.3 Basic electronic components

LEARNING SUMMARY

After studying this section you should be able to:

- *recognise and understand the application of the basic components used in electronic products*

Power supplies

AQA
EDEXCEL
OCR
WJEC
NICCEA

The power for many electronic circuits comes from batteries. There are basically two types of batteries.

Dry batteries

- The standard single cell voltage of a dry battery is **1.5 volts**, see Fig 6.1(a).
- Higher voltages can be achieved by connecting them together in **series**, see Fig 6.1(b).

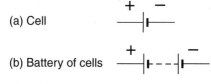

(a) Cell

(b) Battery of cells

Fig 6.1

Advantage	• Portable – can be used to power equipment which is remote from the mains supply
Disadvantages	• Expensive • Must be thrown away when exhausted

Rechargeable batteries

There are two types of rechargeable cell – wet and dry cells.

- The most common example of a wet cell is the **lead acid battery** used in a car. The wet cell battery provides **2.0 volts per cell**.

- The most commonly used dry cell is the **nickel cadmium** cell. This provides only **1.25 volts per cell**.

> **KEY POINT**
> Although initially more expensive than non-rechargeable batteries, their low recharging costs and long life make rechargeable batteries more economic in the long run.

Resistors

> **KEY POINT**
> Resistors resist the flow of electricity through a circuit, i.e. they reduce the amount of current.

- They can have a fixed value, or be variable (known as **potentiometers**).
- The value of a resistor (in ohms) is indicated by a set of coloured bands on its body.

Resistor colour code

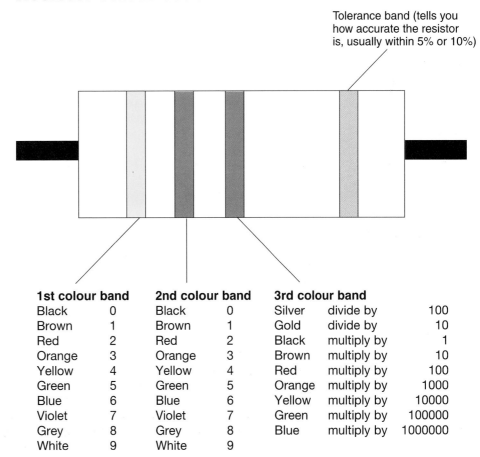

Tolerance band (tells you how accurate the resistor is, usually within 5% or 10%)

Example:
Yellow–violet–brown =
4 : 7 : × 10 = 470
Ohms

1st colour band		2nd colour band		3rd colour band		
Black	0	Black	0	Silver	divide by	100
Brown	1	Brown	1	Gold	divide by	10
Red	2	Red	2	Black	multiply by	1
Orange	3	Orange	3	Brown	multiply by	10
Yellow	4	Yellow	4	Red	multiply by	100
Green	5	Green	5	Orange	multiply by	1000
Blue	6	Blue	6	Yellow	multiply by	10000
Violet	7	Violet	7	Green	multiply by	100000
Grey	8	Grey	8	Blue	multiply by	1000000
White	9	White	9			

Fig 6.2 The resistor colour code

Preferred values

Resistors are commonly available in preferred value series (BS2488).

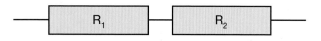
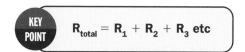

E12 range (10%)	10 12 15 18 22 27 33 39 47 56 68 82
E24 range (5%)	10 11 12 13 15 16 18 20 22 24 27 30 33 36 39 43 47 51 56 62 68 75 82 91

Resistors in series

The total value of resistors in series is simply found by **adding** all the individual values of the resistors in the chain together.

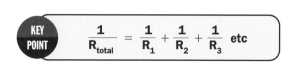

KEY POINT $R_{total} = R_1 + R_2 + R_3$ etc

Fig 6.3 Resistors in series

Resistors in parallel

In order to work out the effective resistance of resistors used in parallel, it is necessary to use **reciprocals**. The reciprocal of the total resistance is equal to the sum of the reciprocals of the individual resistors used.

KEY POINT $\dfrac{1}{R_{total}} = \dfrac{1}{R_1} + \dfrac{1}{R_2} + \dfrac{1}{R_3}$ etc

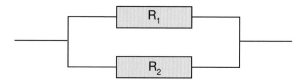

Fig 6.4 Resistors in parallel

Variable resistors

> An example is the volume control on a radio or CD player.

These are known as potentiometers or **'pots'** for short.

- Their resistance is varied by the use of a knob or slider. Variable resistors are often used for changing the **sensitivity** of a sensing device or reducing the feedback to an amplifier.

Light-dependent resistor (LDR)

> An ideal application would be the light sensor for switching circuits.

LDRs are special resistors whose resistance changes according to the **level of light** falling on their surface – as light goes up, resistance goes down.

- The resistance can change from 10 MΩ in the dark to about 1 kΩ in bright light.

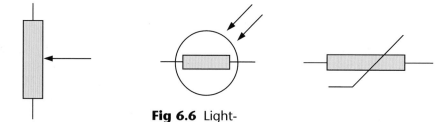

Fig 6.5 Variable resistor

Fig 6.6 Light-dependent resistor

Fig 6.7 Thermistor

Thermistor

> This makes them ideal as temperature sensors.

Thermistors are special resistors whose resistance changes according to the **temperature** – most commonly, as the temperature goes up, the resistance goes down.

Capacitors

AQA
EDEXCEL
OCR
WJEC
NICCEA

> **KEY POINT**
> A capacitor is a device that is able to store electrical energy. The unit of capacitance is the farad.

> It is the ability to charge and discharge which makes the capacitor a basic component in electronic timing circuits.

- When a capacitor is connected across a DC voltage supply, with a current-limiting series resistor, it will charge up to the supply voltage in a **short period of time**.
- Once a capacitor is charged, the current **leaks away slowly**.

There are three types of capacitor.

Variable	Not very common, used mainly for tuning radio circuits	
Fixed electrolytic	A fixed value, which has positive and negative terminals and must therefore be connected the right way round in the circuit	
Fixed non-electrolytic	Can be connected either way round	

Switches

AQA
EDEXCEL
OCR
WJEC
NICCEA

> **KEY POINT**
> Switches are mechanical devices that can make (connect), or break (disconnect) a circuit and can have different numbers of poles and throws.

- **Poles** are the number of connections they can make.
- **Throws** are the number of positions to which each pole can be connected.

Single pole single throw (SPST)	(normally open) / (normally closed)	Single pole double throw (SPDT)	
Double pole single throw (DPST)		Double pole double throw (DPDT)	
Push to make		Push to break	
Rotary		Reed	

Diodes

AQA
EDEXCEL
OCR
WJEC
NICCEA

KEY POINT Diodes are semiconductor devices, which only allow the current to flow in one direction.

There are two ways of connecting a diode in a circuit.

Remember, current flows when the 'arrows' match.

Forward biasing	• The diode is connected the 'right way round', with the positive side (anode) connected to the positive supply • The diode **will conduct**	
Reverse biasing	• The diode is connected the 'wrong way round', with the negative side (cathode) connected to the positive supply • The diode **will not conduct**	

Light-emitting diode (LED)

LEDs are usually used to indicate that part of a circuit is operating.

This type of diode **gives off light** when connected the right way round (forward biasing).

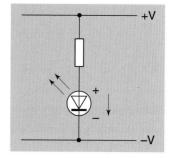

• LEDs can only take a small current, usually **30 mA** maximum, and are therefore usually connected in series with a current-limiting resistor to prevent damage.

Fig 6.8 Light-emitting diode

Thyristors (silicon controlled rectifiers)

AQA
EDEXCEL
OCR
WJEC
NICCEA

KEY POINT In their normal state, thyristors have a very high resistance and do not conduct. However, when they are turned on, they conduct in the forward direction just like a diode.

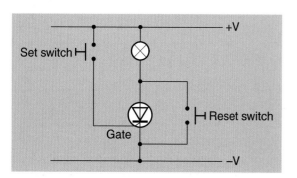

Fig 6.9 Thyristor circuit using a reset switch

• Whilst they are mainly used in AC circuits, they can be useful in DC circuits as a **'latch'**.
• A small current applied to the **gate** of a thyristor will cause it to conduct, remaining 'latched' in this position until the current flowing through it is reduced to a certain minimum.

Relays

AQA
EDEXCEL
OCR
WJEC
NICCEA

KEY POINT A relay is an **electromechanical device** that can be used as an **interface** e.g. between a battery-operated control circuit and a mains voltage operating circuit.

- When current passes through the **coil**, it creates a magnetic field. The coil then acts as a magnet, attracting the **contacts** of the switch, causing it to operate.

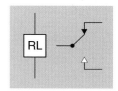

Fig 6.10 Relay

Integrated circuits (ICs)

AQA
EDEXCEL
OCR
WJEC
NICCEA

KEY POINT These are small **electronic circuits** that have many different components built onto a small slice of **silicon** (semiconductor) and housed in a plastic case.

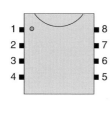

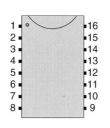

- They are usually set up in a **DIL** (dual in line) configuration, where there are two sets of parallel connections **(pins)**.
- A **small circle** usually identifies pin 1, and the remaining pins are numbered **anticlockwise**.

Fig 6.11 Pin numbers of integrated circuits

Transistors

AQA
EDEXCEL
OCR
WJEC
NICCEA

KEY POINT Bi-polar transistors are three-lead current-operated devices that can provide **amplification**, i.e. the output current is greater than the input current.

> The arrow always points to the negative supply.

- The three leads are called the **base** (b), the **emitter** (e) and the **collector** (c).
- The amount by which the transistor amplifies current is called the **gain** (h_{FE}) of the transistor.

There are two main types of transistor, the n-p-n and the p-n-p.

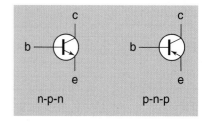

Fig 6.12 Transistors

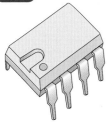

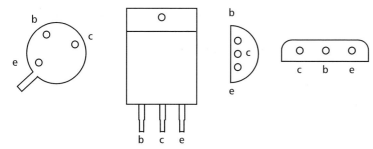

Fig 6.13 Labelling of transistor leads

Identifying the leads

The most common transistor layouts, looking from **underneath**, are shown in Fig 6.13.

The transistor as a current amplifier

When a small current flows through the base-emitter circuit, a much larger current flows through the collector-emitter circuit. This is especially useful in sensing circuits, e.g. when a small current from a light sensor switches a relay to turn on the mains electricity for a porch light.

- As both the base current and the collector current flow through the emitter:
  $$I_e = I_c + I_b$$
- The voltage to the base of the transistor is fed through a current-limiting resistor. A **minimum voltage of 0.6 V–0.8 V** (normally assumed to be 0.7 V) is needed before the transistor will begin to switch – this is known as the **threshold voltage**.

Transistor gain

The current flowing into the base of the transistor is much smaller than the current flowing from the collector to the emitter. The ratio of these currents is called the gain (h_{FE}).

- $$h_{FE} = \frac{I_c}{I_b}$$

> Transistors should be used with a heatsink to reduce the heat when in use and protect them from overheating during soldering.

Transistor power

The power of a transistor is related to the **collector current** that can pass before the transistor overheats and stops working. This rating must not be exceeded.

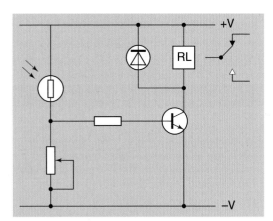

Fig 6.14 The transistor as a switch

The transistor as a switch

The transistor is often used as a **switch**, which swings from the 'off' position to a fully 'open' position when a base current is applied.

- This is very useful in sensing circuits as shown in Fig 6.14 where, as the light level reaches a certain point, the relay will switch on completely.

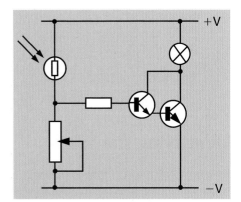

Fig 6.15 Darlington pair

Darlington pair

To increase the sensitivity (increase the **amplification**) it is common to use the Darlington pair arrangement.

- Two transistors are connected together as in Fig 6.15. Because the base current for the second transistor is the emitter current from the first, a **low base current** can trigger a **high collector current**.
- This arrangement acts like a single transistor with its gain equal to the gains of the two transistors **multiplied** together. A high gain and a high output current are achieved.

The field effect transistor (FET)

The FET is a transistor with a high input impedance (resistance), which means that it **draws very little current** from the signal source.

● It is often used in radio and amplifier circuits with small gains but large input impedance.

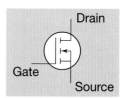

Fig 6.16 Field effect transistor

6.4 Timing circuits

LEARNING SUMMARY

After studying this section you should be able to:

● **create time delays**
● **use the 555 timer IC**
● **recognise the difference between the monostable and astable timer**
● **understand frequency**

Creating time delays

AQA
EDEXCEL
OCR
WJEC
NICCEA

The resistor and the capacitor are the basis of most timer circuits – often referred to as **RC circuits**.

● Its size, and that of the resistor used with it, determines the time a capacitor takes to charge or discharge. The time taken to charge a capacitor to two-thirds of its supply voltage is called the **time constant**.

KEY POINT
The time constant can be calculated by:
t (seconds) = C (farads) × R (ohms)

The 555 timer IC

AQA
EDEXCEL
OCR
WJEC
NICCEA

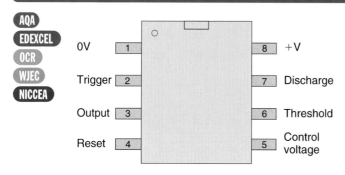

Fig 6.17 NE555 timer IC

● These are cheap to buy and come in an 8-pin DIL pack.
● One of its main advantages is that it can be used both as an astable and monostable timer.

KEY POINT
Although timing circuits can be developed using discrete components (resistors, capacitors, etc), it is more usual to use the 555 timer IC (integrated circuit).

Monostable	● This circuit has only **one stable state** – this means that it remains off (logic 0) until triggered ● When a positive voltage (0 V–5 V) is applied to the trigger, the output goes high (logic 1) for a time determined by the values of the resistors and capacitors – it then returns to its original state until triggered again ● Ideal for devices such as burglar alarms, which sound for a set time once activated	
Astable	● This circuit has **no stable output** – this means that it continually switches from low output (logic 0) to high output (logic 1) ● The astable generates **a pulse**, the time period of which is determined by the values of the resistors and capacitors used ● This makes it ideal for use as a clock	On … Off (waveform)

Frequency of an astable timer

AQA
EDEXCEL
OCR
WJEC
NICCEA

● The frequency is the number of changes (or cycles) in one second.

> **KEY POINT**
> The unit of frequency is the Hertz (Hz).
> 1 Hz = 1 pulse per second

> The clock speed of computers can be very high, e.g. 200 MHz – 200,000,000 pulses per second.

6.5 Amplifiers

> **LEARNING SUMMARY**
> *After studying this section you should be able to:*
> ● *understand the operation of the operational amplifier*
> ● *understand the use of feedback to control output*

The 741 operational amplifier

AQA
EDEXCEL
OCR
WJEC
NICCEA

> **KEY POINT**
> Unlike transistors, which are current amplifiers, the 741 op amp is a voltage amplifier.

It comes in an 8-pin DIL arrangement as shown in Fig 6.18. Its main features are that it:

● has a gain of approximately 100,000
● uses a **split power supply** – requiring a positive voltage, a negative voltage and a 0 V supply

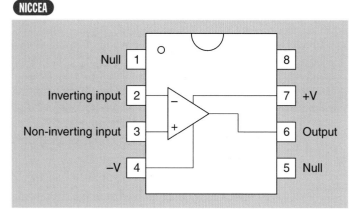

Fig 6.18 741 pin connections

The 741 as a comparator

● When used as a comparator, the 741 will compare two inputs and amplify the difference.
● This makes it very good for use in sensing circuits.

As it has a high impedance (internal resistance), it will only source or sink about 10 mA. Because of this it needs a power transistor connected to the output in order to power output devices such as a motor, lamp or heater, see Fig 6.19.

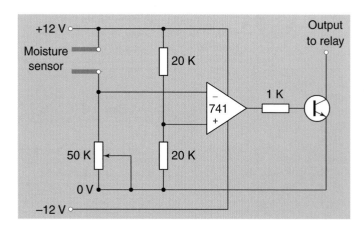

Fig 6.19 Op-amp moisture sensor controlling a watering system

Voltage amplifiers and feedback

AQA
EDEXCEL
OCR
WJEC
NICCEA

One of the most common uses for voltage amplifiers is in audio systems, e.g. amplifying the small electrical signals from the pickups of a guitar so that they can be heard. If the 741 was used as a comparator, it would simply amplify any small differences, producing the maximum output voltage.

Inverting amplifiers

By feeding back some of the output (negative feedback) into the inverting input, the amplification can be controlled, see Fig 6.20(a).

- As the name suggests, this arrangement will **reverse the polarity** of the input.
- $$\text{Gain} = -\frac{R_f}{R_{in}}$$

Note: In Fig 6.20(b), although the input is supplied to the non-inverting input, the feedback is applied to the inverting input.

Non-inverting amplifiers

If the non-inverting input is used, as in Fig 6.20(b), then:

- the polarity of the input will **remain the same**
- $$\text{Gain} = \frac{R_{in} + R_f}{R_{in}}$$

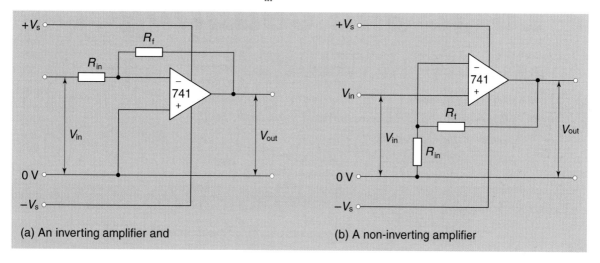

(a) An inverting amplifier and

(b) A non-inverting amplifier

Fig 6.20 Inverting and non-inverting amplifier circuits

6.6 Alternating current (AC)

LEARNING SUMMARY

After studying this section you should be able to:

● *understand the main properties of alternating current*

KEY POINT

An alternating current supply is one that alternates from positive to negative at regular intervals. Mains electricity, for example, is supplied at 240 V, 50 Hz.

There are three important properties to know.

● **Cycle** – one cycle is a complete set of positive to negative values

● **Period** – the time taken in seconds to complete one cycle

● **Frequency** – the number of complete cycles in a one second period

> This is the shape of a sine wave.

Fig 6.21 AC waveform

6.7 Logic and counting

LEARNING SUMMARY

After studying this section you should be able to:

● *use truth tables*
● *understand the difference between the logic families*
● *count using logic circuits*
● *display the output*

Truth tables

AQA
EDEXCEL
OCR
WJEC
NICCEA

Digital circuits are built from simple on/off switches called **gates**. These gates have **two states**.

● On, or logic high (1)
● Off, or logic low (0)

KEY POINT

Logic, or truth tables, **can be produced to analyse clearly the possible alternative states of a digital circuit.**

> The truth table for this circuit is called an AND gate.

For example, in Fig 6.22, the output will only be high (logic 1) when **both** switches A and B are on.

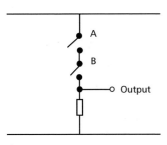

Fig 6.22

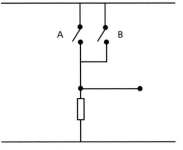

Fig 6.23

> The truth table for this circuit is called an OR gate.

However, in Fig 6.23, the output will go high if **either** switch A or B is on.

All logic circuits can be made from switches connected together in order to achieve desired output states. There are **six types of logic gates** used.

Gate	Symbol	Truth table
AND		A B Output 0 0 0 0 1 0 1 0 0 1 1 1
OR		A B Output 0 0 0 0 1 1 1 0 1 1 1 1
NOT (or inverter)		A Output 0 1 1 0
NAND (an inverted AND)		A B Output 0 0 1 0 1 1 1 0 1 1 1 0
NOR (an inverted OR)		A B Output 0 0 1 0 1 0 1 0 0 1 1 0
Exclusive OR (XOR)		A B Output 0 0 0 0 1 1 1 0 1 1 1 0 (only logic 1 when **either** input is 1 **not** when both)

Logic families

There are two distinct logic families.

- **Transistor transistor logic (TTL)** – these ICs are listed as the 7400 series.
- **Complementary metal oxide semiconductors (CMOS)** – listed as the 4000B series.

Property	TTL	CMOS
Power supply	5 V (±0.25 V)	3–15 V
Current required	3 mA	8 μA
Input impedance	Low	High
Temperature range	0–70°C	−40 to +85°C
Operating speed	Fast	Relatively slow

Counting with logic circuits

AQA
EDEXCEL
OCR
WJEC
NICCEA

> **KEY POINT** The simplest counter is a bistable or flip-flop.

- The example in Fig 6.24 uses NOR gates.
- The feedback latches the output until the reset switch is pressed.

S-R flip-flops are mainly used for **simple latches**, e.g. operating a solenoid driven bolt on an automatic locking device – holding it open until the reset switch is pressed. However, this type of latch has the disadvantage of having no control over when it operates after the set switch has been pressed.

> This type of bistable is called an S-R flip-flop, standing for Set–Reset.

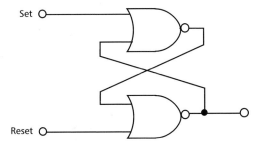

Fig 6.24 S-R flip-flop

> A register could be taken using D-type flip-flops, counting inputs only when a master switch (ST) is pressed.

- An improvement is to use a **clocked bistable**, such as the **D-type flip-flop**, see Fig 6.25. This has the advantage that it is triggered by a **rising edge** (increasing voltage), which can be controlled, rather than a complete pulse (0–1).

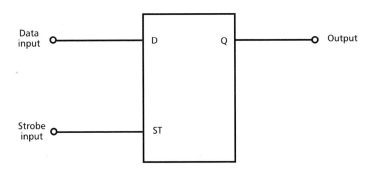

Fig 6.25 D-type flip-flop

Displaying the output

AQA
EDEXCEL
OCR
WJEC
NICCEA

When using a D-type flip-flop, one good way of displaying the output is to use an **LED**.

Displaying numbers

> **KEY POINT** However, the most common way to display numbers is to use the seven-segment display.

There are basically two types of display.

- **Liquid crystal display (LCD)** – uses virtually no current, working on reflected light and is therefore **ineffective in the dark**.
- **LED display** – draws relatively large amounts of current, but **will work in the dark**.

A lower-case letter is used to identify each segment of the display, see Fig 6.26.

- Each segment can be switched on and off independently in order to create the numbers 0–9.
- As can be seen in Fig 6.26, the number 2 is created by using only segments a, b, d, e, and g.

> Only a limited selection of the alphabet can be created using this form of display.

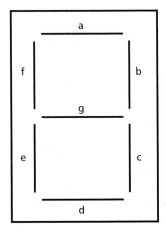
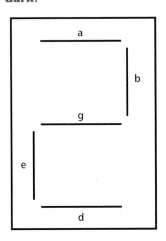

Fig 6.26 Seven-segment display

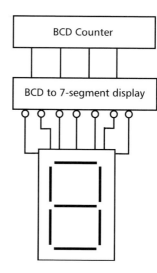

Fig 6.27 BCD and seven-segment display arrangement

In order to convert the output from a counter into a number display, the **binary output** (a 'word' made up of ones and zeros) must be decoded.

- A **binary coded decimal** (BCD) chip is used.
- The BCD receives a binary code and converts this to the seven-segment display format.

Debouncing

It is very important when designing any counting circuits that a **smooth signal** comes from the input switch.

- Mechanical switches suffer from the problem of '**bounce**' in that, when they are pressed, the contacts bounce off each other several times before they close resulting in a '**false count**'.
- The S-R flip-flop can be used to '**debounce**' the switch signal as shown in Fig 6.28.

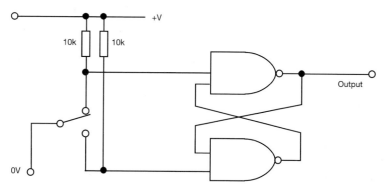

Fig 6.28 Debounce circuit

6.8 Pneumatics

After studying this section you should be able to:

- **recognise components and their symbols**
- **understand their potential for use in control systems**

Pneumatic components

AQA
EDEXCEL
OCR
WJEC
NICCEA

The following are the BS/ISO standard symbols and conventions used when constructing pneumatic circuits.

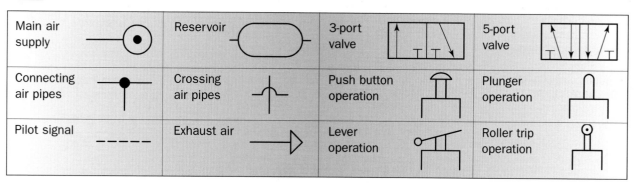

Main air supply		Reservoir		3-port valve		5-port valve	
Connecting air pipes		Crossing air pipes		Push button operation		Plunger operation	
Pilot signal		Exhaust air		Lever operation		Roller trip operation	

Pressure sensitive/ diaphragm operation		Solenoid operation		Air bleed		Spring return	
Single-acting cylinder		Double-acting cylinder		Flow regulator/ unidirectional flow control valve		Shuttle valve	

Control operations

AQA
EDEXCEL
OCR
WJEC
NICCEA

Pneumatic circuits can replicate many of the control operations achieved by electronic circuitry. You should make yourself familiar with the following applications, especially with regard to **sequential control** and **automatic reciprocation**.

Time delays

> Pneumatic time delays are often used in safety systems.

- These can be achieved by the use of a **reservoir** or a unidirectional flow control valve (**restrictor**).
- Because air is always being lost during operation, **pressure fluctuations** can affect the **accuracy** of such time delay circuits.

Signal amplification

> Air bleed systems tend to be used for sensing and counting.

- **Feedback**, the basis for amplification, can be generated by the use of an **air bleed**.

Logic functions

> An example of an OR gate would be the ability to open a sliding door from both sides.

- **OR gates** can be achieved by using a **shuttle valve**.
- Simple **AND functions** can be achieved by using **two 3-port valves**.

> **KEY POINT**
> Safety circuits are a common application for logic circuits when pneumatics control circuits are used, especially those requiring multi-point operation.

Force/movement from a pneumatic cylinder

Compressed air supplied under pressure

Surface area of piston

Piston rod

Spring

Cylinder

The force and magnitude of linear movement produced by a pneumatic cylinder depends on three elements.

- The **pressure** at which the compressed air is supplied (N/mm²)
- The **surface area** of the piston (mm²)
- The clear **length** of the cylinder (mm)

> A double-acting cylinder produces different forces on each stroke due to the effect of the piston rod in reducing surface area on one side of the piston.

Fig 6.29 Single-acting pneumatic cylinder

> **KEY POINT**
> Using the fact: Force = Pressure × Area, then
> Force produced by a cylinder = air pressure × piston area
> (N)   (N/mm²)   (mm²)

Textiles technnology

The following topics are covered in this chapter:

- Equipment
- Materials
- Working properties
- Colouring fabrics
- Decoration
- Construction techniques
- Fabric care

7.1 Equipment

LEARNING SUMMARY

After studying this section you should be able to:

- select, name and safely use pieces of equipment for a particular task or process

Applications of equipment

AQA
EDEXCEL
OCR
WJEC
NICCEA

Whilst it is likely that you will only have direct experience of working in a domestic/school environment, you should also understand the applications on an industrial scale. This will involve the scaling up of your product for manufacture in terms of quantities and using industrial sized equipment.

Equipment	Comments	Appearance
Scissors and shears	• Shears are used for cutting out the fabric, whilst the smaller scissors are required for trimming and cutting thread • These scissors must not used on paper, as it will blunt them quickly	
Pinking shears	• Used for finishing seams on fabrics that do not fray	
Textile scissors	• Used for snipping off thread ends	
Tailor's chalk	• Used for marking out accurately prior to cutting – especially on darker fabrics	
Tracing wheel	• Used with dressmaker's carbon for transferring markings	

Equipment	Comments	Appearance
Tape and metre rule	• A tape is used for taking measurements, while a rule is used for measuring the fabric itself	
Paper patterns	• Made from thin tissue, they are used as a template to cut fabric to the required shape • Also contain information to assist construction	
Pins	• Used to temporarily hold fabric together during construction, or to mark positions	
Thimbles	• Protect the fingers when hand sewing, especially for long periods, or when using tough/thick fabrics	
Needles	• Used for sewing and embroidery. It is important to select the correct size and type for the task • Types include: betweens, milliner's or straw, crewel, tapestry and chenille	
Machine needles	• Specially designed for working with sewing machines. The selection of the correct needle for the fabric and thread is important	
Stitch unpicker	• Used to unpick seams to avoid damaging the fabric	
Bodkin	• Made from metal or plastic, it is used for threading cord or elastic through hems and casings	
Embroidery frame	• Maintains the tension of the warp and weft threads in a fabric by enclosing it in two sprung rings	
Iron – dry or steam	• Used to remove unwanted creases or to deliberately form a crease in a fabric • Available in various weights	
Ironing board	• Used to place fabrics or garments on when pressing or ironing	
Sleeve board	• Should have a velvet pressing board available as well for use with pile fabrics	

Equipment	Comments Appearance	
Knitting machine	• Can be hand operated or controlled automatically by punch card or buttons • A computer, as part of a CAD/CAM process, can be used to control more modern machines	
Overlocker	• Used to finish off cut edges of fabrics • They are also used to construct knitwear, producing a seam that can stretch with the fabric	See Fig 7.1

The sewing machine

AQA
EDEXCEL
OCR
WJEC
NICCEA

Your understanding of how the sewing machine works could be extremely important in terms of the quality of your coursework. You are advised to practise using it whenever possible.

The sewing machine is the 'workhorse' of textiles technology. A wide range of **stitch selections** is available and some machines are designed to carry out **specialist tasks**, e.g. overlocker.

• Many modern machines are **programmable**, with the more sophisticated ones being **computer controlled**.

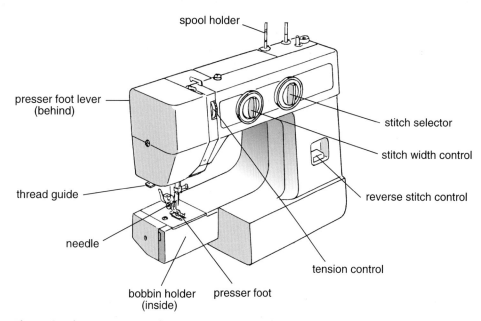

Fig 7.1 The sewing machine

– – – – – – – – – a small straight stitch

— — — — — a long straight stitch

∧∧∧∧∧ a zig-zag stitch

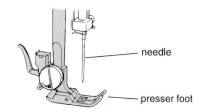

Fig 7.1 The sewing machine (continued)

7.2 Materials

After studying this section you should be able to:

● *recognise the difference between various fibres*
● *understand how yarns are produced*
● *know how fabrics are constructed*

Fibres and their properties

AQA
EDEXCEL
OCR
WJEC
NICCEA

Textiles use **fabrics** that are made out of **fibres**. The fibres can be short in length (known as **staple fibres**), requiring spinning into a yarn before use. They can also be long (known as **filaments**), and are used as they are or chopped into short lengths and spun into a yarn.

Filament fibres can also be **processed** further to increase their bulk by introducing small loops, curls and twists which produce thicker, springier yarns.

KEY POINT

You should be aware of the technological developments made in textiles materials, especially in 'smart' fibres and fabrics, and understand their uses in various industries.

> You are advised to collect samples of various fabrics noting their appearance, texture, uses and working properties.

There are two distinct types of fibre.

Natural	● Animal
	  ● Silk – silkworm
	  ● Wool – sheep
	  ● Hair – camel, goat (mohair, cashmere), horse, rabbit (angora)
	● Vegetable
	  ● Seed – cotton, kapok, coir
	  ● Bast – flax, hemp, jute
	  ● Leaf – manila, sisal
	● Mineral
	  ● Asbestos

Synthetic (man-made)	● Synthetic polymer – polyvinyl, polyurethane, polyester, polyamide (nylon) ● Natural polymer ● Rubber, regenerated protein and cellulose, cellulose ● Other – carbon, glass fibre, ceramic, metal

> You should be able to recognise the origin of a wide range of fibres.

Natural fibres

Cotton	● **Staple fibre** made from the fibres surrounding the seed head of the plant, producing **durable** fabrics – especially the thicker varieties ● Texture of the fabric is **smooth and firm** and it is **easy to sew** ● **Creases badly**, having poor elasticity, but washes well ● Accepts **dyes well**, but lacks lustre unless treated ● **Burns quickly** with the smell of burning paper, remaining alight out of the flame ● Suitable for **light clothing and bed linen**, especially when **combined with polyester** to reduce creasing
Wool	● **Staple fibre** with a natural crimp that produces hard-wearing fabric which can be **firm to very soft** in texture ● Absorbs water but **shrinks badly**, requiring damp pressing ● Naturally **elastic**, it presses back well into shape ● Will **shed natural creases** well, whilst still being able to **hold pressed creases** ● Accepts **dyes very well** ● **Smoulders** in a flame with the smell of burning hair, but extinguishes itself out of the flame ● Depending on the weave, it is reasonably easy to sew, producing **warm clothing**. It is also suitable for **soft furnishings and carpets**

Linen	<ul><li>**Staple fibre** made from the stems of the **flax plant** that is strong and hard wearing, easy to sew and good for producing thick fabrics that have a **firm to rough** texture</li><li>**Absorbs water** well, but creases easily and therefore **needs ironing**, which gives it a **good lustre**</li><li>**Accepts dyes** very well</li><li>Burns in a similar manner to cotton, but with a **flame that flares**</li><li>Increasingly **expensive** to produce, but is suitable for **tea towels, tablecloths and napkins**</li></ul>
Silk	<ul><li>**Filament fibre** produced by unravelling the cocoon of the **silkworm** in water. It is a very fine filament that can be used as it is, but also can be chopped and spun or processed to increase its bulk by adding loops and twists to give a thicker and even more elastic yarn</li><li>The finer weaves need careful handling and, although it is **reasonably crease resistant** and good in water, it needs ironing</li><li>**Accepts dyes** well</li><li>**Burns slowly** with a yellow flame and a smell like burning hair, but goes out when not in the flame</li><li>Can produce a **strong, hard-wearing fabric** that is **smooth to soft** in texture, with an **excellent sheen** and good thermal properties</li></ul>

Synthetic fibres

Polyester	<ul><li>**Filament or staple fibre** made from ethylene glycol and terephthalic acid</li><li>Has **poor natural elasticity**, but the spun fibres can be crimped to improve this (Crimplene)</li><li>Does **not dye well** at home</li><li>Burns with a **sooty flame**, melting and shrinking, and self extinguishes</li><li>As it **washes well**, drips dry and resists creasing, it is often **mixed with cotton** for shirts and blouses</li><li>Can also be **woven into a heavier** fabric for suits, and it holds a pressed crease very well, although can tend to **build up static**</li></ul>
Acrylic	<ul><li>**Filament fibre** made from oil, but more often used as a **staple fibre knitted yarn** form to produce an inexpensive **'wool substitute'** that wears well</li><li>Has **poor thermal** properties when used on its own and **builds up static**</li><li>Does **not absorb water** and is therefore **poor for dyeing**, but drips dry</li><li>Burns quickly with a **spluttering flame** that stays alight out of the flame, and an **acrid smell**</li></ul>
Polyamide (nylon)	<ul><li>Produced as **filament and staple fibres** from chemicals derived from oil and coal</li><li>**Smooth, cold** fibre that has much improved thermal properties in its **'brushed'** form</li><li>Strong and hard-wearing, but **builds up static** readily</li><li>Washes and dries very well, being extremely **crease resistant**</li><li>**Takes a dye** reasonably well</li><li>**Melts** rather than burns, producing **droplets** that carry a blue/yellow flame</li><li>Is fairly **difficult to sew**, generally used for shirts, tights, stockings, carpets and waterproof clothing</li></ul>
Polypropylene	<ul><li>**Filament fibre** synthesised from chemicals, producing a very strong and hard-wearing yarn that is **non-absorbent** and fairly **slippery**</li><li>Burns with a **yellow/blue flame** and little smoke, producing **droplets** that go out and smell of candle wax</li><li>Used for ropes, carpets and furnishings</li></ul>
Viscose	<ul><li>**Filament and staple fibre** made from cellulose obtained from wood pulp and is spun wet. It is also known as a **'regenerated fibre'**</li><li>Care needs to be taken when sewing and washing as seams may pull</li><li>**Burns continuously** with a yellow flame</li><li>**Inexpensive**, shiny and not particularly durable fabric that is used for linings</li></ul>
Acetate	<ul><li>Another **regenerated filament and staple fibre** made from wood pulp</li><li>Care is needed when sewing but it is **fairly durable**</li><li>**Burns easily** with a yellow flame, cracking and shrivelling, giving off the smell of vinegar</li><li>Is spun dry to produce a smooth, slippery and shiny fabric that is used for linings of suits and skirts</li></ul>

Lyocell (Tencel)	• A recent development of the cellulose process that is **environmentally friendly** in that it uses renewable wood pulp and a 'closed loop' spinning process that recycles its non-toxic solvent
	• Fibres are **absorbent** and **hold dyes well**
	• Used as a cotton **'substitute'** – being more crease resistant – used for protective clothing and some medical applications
Elastane (Lycra)	• **Filament fibre** that is strong and smooth
	• Does **not take** a dye very well
	• Has **good elasticity** making it a less perishable substitute for rubber
	• Very **poor absorbency** makes it good for drying, and it is used in swimwear, underwear and interlinings

> You should consider properties such as flammability when evaluating the safety of an application.

> **KEY POINT**
> You should revise carefully the common properties and applications of two natural and synthetic fibres, along with the advantages and disadvantages of two fibre combinations, including polyester/cotton.

Microfibres

To achieve fibres that are **extremely fine**, polymers can be mixed to produce microfibres. **Polyester** is usually used, but nylon and acrylic can also be used. They can also be **blended** with other fibres, cotton being a common example.

- Microfibres can be **woven or knitted**, producing fabrics that are lightweight, durable and soft to wear.
- Because they can be woven so tightly, the fabrics can be almost **impermeable** to water yet will let **perspiration escape**. For this reason they are extremely useful for **outdoor** wear.

Smart fibres

As the technology of fibre production has developed, the concept of the **'smart material'** has evolved.

- This essentially is a fabric that alters its properties in a beneficial way, in **response to changes** in heat or light for example.

Yarns

AQA
EDEXCEL
OCR
WJEC
NICCEA

Spinning fibres and filaments makes a yarn – **a single strand** of the material.

> The properties of a fabric will depend on the yarn used.

wool fibres drawn out between teeth of carders

Fig 7.2 Carding by hand

Carding

Staple fibres, such as wool, need to be separated as they are too matted to be used as they are. Passing two wire brushes (carders) across the fibres in opposite directions to pull the fibres apart does this.

This can also be done on an industrial scale by **machine carding**, using the same principle.

Spinning

The spinning (or **twist**) binds the fibres together and produces yarns with differing degrees of strength and softness.

- Clockwise spinning gives an S-twist, whilst anticlockwise spinning produces a Z-twist.
- Single yarns can then be twisted together to produce yarns with **enhanced properties** – thicker or stronger, e.g. 2, 3 or 4 ply.
- Twisting together different types of yarns can produce '**fancy yarns**' with interesting effects, e.g. bouclé and knop yarns.

> **KEY POINT**
> Bulky yarns required for knitting have a long twist, but thinner fabrics require a yarn with more twist to increase its strength.

Fabrics

AQA
EDEXCEL
OCR
WJEC
NICCEA

Fabric is constructed in a number of ways using yarns, fibres (non-woven) or a combination of both.

Weaving

A loom is used to hold lengths of yarn in position. Another yarn is then interwoven through these lengths and pressed home.

- The **longitudinal** yarns are called the **warp**.
- The **cross** yarns are called the **weft**.
- The edge yarns are often stronger, or doubled up, to provide a good edge. This is known as the **selvedge**.

The loom can also be made to adjust the **tension** of the warp and weft, and the **density** of the finished fabric.

When each weft yarn is interwoven with alternate warp yarn, it is known as plain weave.

Different **patterns and texture** can be introduced by changing the colours of both the warp and weft yarns and also by adjusting the number and location of the warp yarns **missed out** by the interweaving of the weft yarn.

- The range of effects produced by the latter methods have names such as **twill, hopsack and satin**.

> **KEY POINT**
> Woven fabrics will stretch more across the weft than the warp. However, they stretch most at an angle across the warp and the weft.

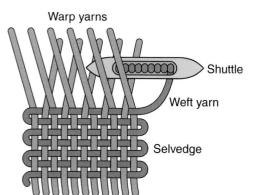

Warp yarns
Shuttle
Weft yarn
Selvedge

Fig 7.3 Weaving using a simple loom

Knitting

Knitting by hand is a slow process, but is considered a satisfying and rewarding pastime and allows extremely creative effects to be produced. However, the process can be readily mechanised for commercial purposes.

The air pockets created give knitted fabrics good thermal properties.

- The basic process is to form the fabric by **interlocking loops of yarn**. Altering the thickness and type of the yarn, the size of the needles (and therefore the size of the loops formed) and the combination of the stitches used can produce the desired texture and pattern changes.

Weft knitting	• The loops run across the width of the fabric or around the circle • Can be done by hand using a single yarn • Will unravel if the yarn is broken	
Warp knitting	• The loops run in a vertical direction • Uses a number of yarns and cannot be produced by hand • Less prone to laddering • The quickest of the two methods	

KEY POINT — **Knitted fabrics** stretch in all directions, but will return to shape if not overstretched or stretched in the same way on a regular basis.

Fig 7.4 Crocheting

Crocheting

This is a hand technique used for producing intricate and decorative fabrics and edgings.

- The yarn is looped with the use of a **crochet hook**, with the chains produced being **interlinked** to produce the desired patterns.
- The general effect is one of a **fabric with 'holes'**.

Macramé

This is another hand technique producing a product with holes.

Lace and nets are other examples of openwork fabrics.

Fig 7.5 Macramé

- Much **thicker yarns**, such as plaited yarns and strips of fabric, string, cord and ropes tend to be used.
- The yarns are **knotted together** (commonly using the square knot and clove hitch) to produce strong fabrics that can be used as hammocks, hangers, rugs and wallhangings.

Felting

This is a non-woven fabric that is cheap and quick to make, the fibres being formed into the fabric by a combination of **moisture, heat and pressure**.

- Felting is used for **carpet underlays** as no real strength or wearing properties are required.
- It is also used for **hats** because the desired shape can be achieved **without seams**.

Bonding

This is another non-woven fabric, where the fibres are bonded together by a **resin adhesive** or **stitching by thread**.

- Bonding is used regularly as **insulating linings** or **disposable cloths** because of good absorbency.

Laminating

Two or more fabrics are stuck or bonded together to combine their properties.

- This is often done to improve the aesthetic qualities of a synthetic fabric.

> **KEY POINT**
> You should investigate thoroughly, by disassembly, the construction of two woven and knitted fabrics and one felted, bonded or laminated fabric.

7.3 *Working properties*

>
> **LEARNING SUMMARY**
>
> *After studying this section you should be able to:*
> - *discuss the physical properties considered when selecting a fabric*
> - *understand the range of applied finishes that can be used*
> - *appreciate the advantages of combining materials to enhance the properties*

Physical properties

AQA
EDEXCEL
OCR
WJEC
NICCEA

A fabric is selected for a particular application based on a number of physical properties.

Washability	Its need to be hand washed, machined washed or dry cleaned, and its resistance to shrinking when wet
Thermal insulation	Its ability to retain or lose heat
Stain resistance	The ease with which spills and marks can be removed
Wear	Its resistance to abrasion and tearing in normal use
Elasticity	Its capacity to stretch and return to its original shape under working conditions
Absorbency	How well it absorbs and retains liquids
Water resistance	How well it prevents the flow of water
Wind resistance	How well it prevents the passage of air
Fire resistance	Its combustibility and the toxicity of the fumes given off
Machinability	The ease with which it can be sewn and made up

It would be advantageous to include the results of testing in your coursework folio.

> **KEY POINT**
> It is advisable to know the simple tests for the properties of fabrics for at least flammability, elasticity and washability.

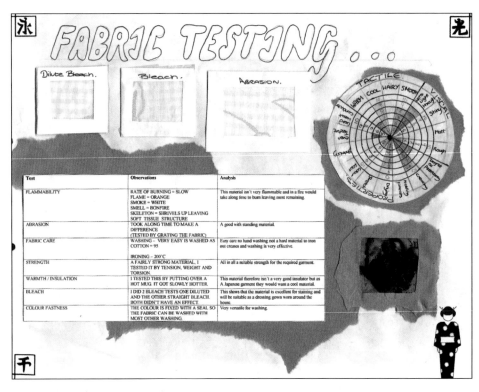

A number of other factors will also affect the choice of fabric.

- Cost and availability
- Colour range
- Weight
- User requirements
- Aesthetics

Fig 7.6 Example of testing used in coursework

Applied surface finishes

AQA
EDEXCEL
OCR
WJEC
NICCEA

The working properties of fabrics can be **improved** by the application of **chemicals** to enhance performance. These finishes may have their **effectiveness reduced** if the fabric is incorrectly treated in **aftercare**, see Section 7.7 on fabric care.

These are some of the surface treatments which can be applied.

Anti-bacterial	Checks the growth of bacteria and its visual effect on the fabric
Anti-static	To reduce the build-up of static electricity by retaining the moisture which conducts it
Crease resistance	Application of resins can reduce the tendency to crease, but can also reduce the fabric's ability to resist wear
Flame retarding	Used to make ignition difficult or to prevent the spread of flame once the source has been removed
Mothproofing	Applied mainly to carpets, upholstery and soft furnishings, making the fibres 'taste' unpleasant
Shower- and waterproofing	Helps the water run off the fabric rather than be absorbed
Stain and soil resistance	A mixture of silicone and fluorine applied to carpets and upholstery

> It is important to research the safety aspects of these applied finishes in your coursework.

Combination

AQA
EDEXCEL
OCR
WJEC
NICCEA

Fabrics are often combined to improve **performance** and **appearance**.

- **Quilting** – for improved insulation
- **Interfacing** – for improved strength or stiffness
- **Reversible fabrics** – to allow dual-purpose use

7.4 Colouring fabrics

LEARNING SUMMARY

After studying this section you should be able to:

- *know why dyes are used*
- *describe how fabrics are dyed*
- *discuss the different painting effects that can be achieved*
- *understand how fabrics are printed*

KEY POINT

Most fibres, synthetic or natural, do not come in a range of colours. Indeed, natural fibres will often be bleached to improve their whiteness.

Dyes

AQA
EDEXCEL
OCR
WJEC
NICCEA

Some **synthetic yarns** can have a **colour pigment** added at the manufacturing stage, but the remaining yarns will need to be **dyed**.

- The fabric can also be **dyed after construction**.

Natural dyes are made from plants, animals and minerals but can be **less fast** and **resistant to fading** than **synthetic dyes** which are chemically produced in a wide range of colours.

KEY POINT

A dye is said to be colour fast if it does not fade during washing. Exposure to light is another common cause of colours fading.

> You should include tests for fastness of dyes in your coursework.

Different fibres react in different ways to different dyes, with **natural fibres** being the best since they are generally more **absorbent**.

- **Synthetic fibres** will not readily accept dyes and require chemicals called **mordants** to **fix** the dyes.

Dyeing fabrics

KEY POINT

The basic principle of dyeing is to immerse the fabric in the dye – often after wetting it first to aid penetration of the dye – rinse out the excess and dry.

When dyeing **bought fabrics** it is often advisable to thoroughly **wash** them to remove any surface treatments that may have been applied that could **inhibit** the dyeing process.

There are a number of methods of dyeing that can be carried out on a small scale which give very effective results, basically relying on **controlling** where the **dye comes into contact** with the fabric.

Batik

A random, 'crazed' effect can be achieved if the waxed fabric is first lightly crumpled, as the wax will crack and allow some penetration of dye.

- **Hot wax** is applied to the fabric – either by using a paintbrush or more traditionally by a **tjanting tool** – so that it saturates the areas that you **do not want to colour**.
- When dry, the wax will **resist** the dye when the fabric is immersed and only the unwaxed areas will be dyed.
- Once the dye has dried, the wax is removed by **hot ironing** between absorbent sheets of paper.

KEY POINT The fact that wax melts when heated indicates that only cold water dyes are suitable for this process.

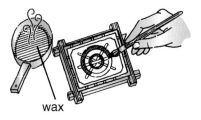

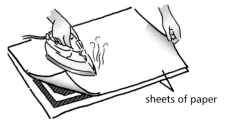

wax

tjanting tool

dye

sheets of paper

Fig 7.7 Batik dyeing

Tie dyeing

Further effects can be achieved by including small pebbles in knots of the fabric.

- This method relies on **tying string, cord or rubber bands** tightly round bundles of the fabric which then resist the dye in interesting and effective ways.
- The process can be **repeated** a number of times with different coloured dyes.

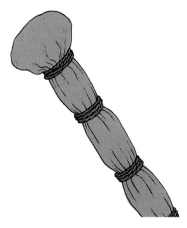

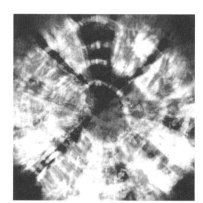

Fig 7.8 Tie dyeing

Tritik

- This is similar to tie dyeing in that the fabric is **pleated and sewn** in order that the dye is resisted in some areas.

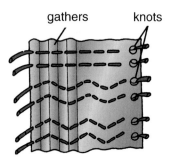

gathers   knots

Fig 7.9 Tritik dyeing

Painting fabrics

AQA
EDEXCEL
OCR
WJEC
NICCEA

As with all transfer methods, the original artwork must be a mirror image of the desired outcome.

Fabric pens and crayons

This is an extremely quick and easy way of applying coloured decoration and is ideally suited for **one-off** items and **mock-ups**.

- They are **waterproof** and available in a wide range of colours.
- **Transfer crayons** are also available, the design being transferred from paper to the fabric by using a hot iron.

Brush painting

The simplest way of applying **fabric paints** is by a **small, soft brush**. In this way, quite detailed decoration can be applied.

- As the paint needs to be **thinned**, it is most suitable for light-coloured, thin fabrics such as silk.

Spray painting

Fabric paints can also be applied using a **spray diffuser** or an **airbrush**.

- The best results are obtained by using a **pre-cut stencil** to **mask** the areas not to be sprayed.

> **KEY POINT** Fabric paints usually need to be set by hot ironing through a sheet of paper.

Marbling

An interesting and impressive effect, resembling colourful marble, can be achieved by **floating the paint** on the surface of a bath of water containing **wallpaper paste**.

- The fabric is **pulled through** the paint after it has been **randomly stirred** to create the marbling effect.
- **Oil-based** paints, as well as fabric paints, are suitable for this technique.

Printing fabrics

AQA
EDEXCEL
OCR
WJEC
NICCEA

Blocks can even be made from vegetables such as potatoes or large carrots.

For producing repeat patterns, printing methods need to be used.

For more detail of printing methods, see Chapter 4, Graphic products.

Block printing

- The desired pattern is formed in **relief** on the surface of a flat block of wood or lino.
- Patterns can also be made by attaching a number of suitable materials, such as card, rubber or string, to a flat block of wood.
- The surface of the block then has the paint applied by **roller** and the block is then **pressed** onto the fabric, which has been washed and pressed flat.
- The block is **repainted** and the process repeated to cover the required area.

> **KEY POINT** Different colours may be subsequently applied on different blocks to create more intricate patterns, but this requires careful registration or indexing to avoid unwanted overlapping.

Screen-printing

- A nylon or polyester fabric is stretched over a frame, through which the paint can be pushed using a **squeegee**, producing a very even application of paint with accurate positioning.

- The fabric to be printed is placed on a flat surface and a **stencil** made from card or thin plastic is put in position.

- The screen is then placed onto this and the paint applied.

> **KEY POINT** A more effective method of screen-printing is when the mesh is treated with a **resist medium** to make the stencil unnecessary. This can be done **photographically**.

Roller printing

Commercially, if large runs of fabric are to be printed, the relief patterns will be put onto rollers, which are inked as the fabric is continuously run under them.

- This can be achieved on a small scale by the use of a hand roller with the pattern attached.

> **KEY POINT** You should investigate the methods used for **registration** as part of quality control.

Transfer papers

 You can produce your own transfer design by using ink jet transfer papers.

- These can be used to enhance projects by allowing prepared images to be ironed onto the fabric, and are often used commercially for small-run **T-shirt printing**.

7.5 Decoration

> **LEARNING SUMMARY** *After studying this section you should be able to:*
> - *select and use appropriate methods of applied decoration*

The use of coloured yarns and fabrics is not the only means of improving the aesthetic qualities of textile products.

Methods of decoration

AQA
EDEXCEL
OCR
WJEC
NICCEA

Quilting

A layer of **polyester wadding** is placed between **two layers** of fabric and sewn together.

- The sewing can be at the edge, across in a regular pattern – a diamond for example – or to produce a more creative effect.

> **KEY POINT** The resulting 'bulging' is not only decorative, but has the advantage that as air is trapped in the layers and pockets formed, the thermal insulation properties are markedly improved.

Appliqué

This is simply the sewing onto one fabric, shapes cut from other fabrics to create **decoration and illustration**.

> To improve the relief effect, wadding is sometimes included.

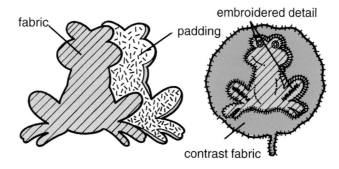

Fig 7.10 Appliqué

Patchwork

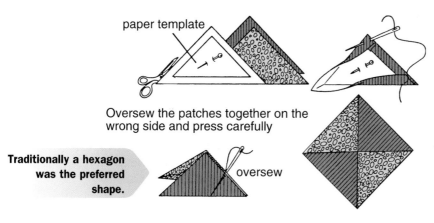

> Traditionally a hexagon was the preferred shape.

Fig 7.11 Patchwork decoration

This is the creation of one fabric by **sewing together scraps** of other fabrics using shapes that **tessellate**.

- The pieces of fabric are cut out using a **template**, including a **seam allowance**.
- These pieces are then systematically and accurately sewn together, allowing for any design that is required to be incorporated within the shapes by selected use of **colour, pattern and texture**.

Embroidery

Coloured threads and yarns being sewn onto the fabric using a wide variety of **stitches in combination** can achieve further decorative effects, often very detailed in design.

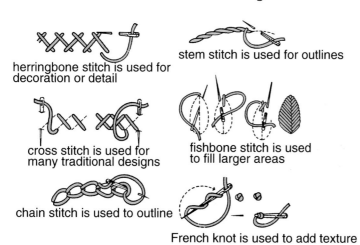

herringbone stitch is used for decoration or detail

stem stitch is used for outlines

cross stitch is used for many traditional designs

fishbone stitch is used to fill larger areas

chain stitch is used to outline

French knot is used to add texture

- **Commercially**, programmable machines produce logos, badges and motifs, as well as labels.

> **KEY POINT**
> If available, it would be advantageous to use the CAD/CAM opportunities offered by a computer-controlled sewing machine to design and produce decorative effects in your coursework.

Fig 7.12 Examples of embroidery stitches

7.6 Construction techniques

After studying this section you should be able to:

● **cut the fabric out using a pattern**
● **select the appropriate seam or hem**
● **understand the uses of various fastenings**

Patterns

Patterns allow a **flat** piece of fabric to be cut so that it will create a **three-dimensional** form when the various pieces are joined.

KEY POINT

You can create patterns from new or adjust existing patterns. Most patterns are produced for basic styles and average sizes (block patterns) – adjustments must be made accordingly.

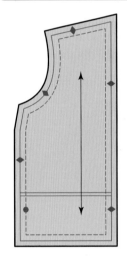

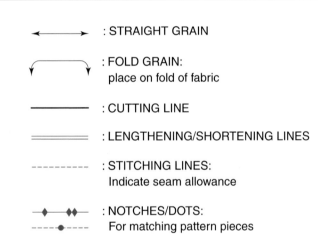

: STRAIGHT GRAIN

: FOLD GRAIN:
place on fold of fabric

: CUTTING LINE

: LENGTHENING/SHORTENING LINES

: STITCHING LINES:
Indicate seam allowance

: NOTCHES/DOTS:
For matching pattern pieces

Fig 7.13 Pattern details

Seams and hems

Plain seam	A simple row of stitches is used to join two pieces of fabric together. The fabric is pinned or tacked right sides together so that the 'spare' material can be pressed flat on the unseen side	
Double-stitched seam	A more decorative form of double seam	fold over

French seam	A double seam used for fabrics that fray easily. The fabric is stitched with the wrong sides together, but the cut edges are protected within the fold.	
Top-stitched seam	A plain seam with extra stitching for strength	
Fusible web	A glue-impregnated web which is activated by heat is ironed inside the folded seam to avoid the use of visible stitches.	
Slip hemming	Hemming using very small stitches on the inside, which are effectively invisible	

Fastenings

AQA
EDEXCEL
OCR
WJEC
NICCEA

Buttons and buttonholes

- Buttons are produced in a vast range of shapes, sizes, styles and material finishes and therefore can be considered to be another form of **applied decoration**.

- They are generally capable of being attached by **machine** as well as by hand.

- Buttonholes can easily be made using a standard sewing machine.

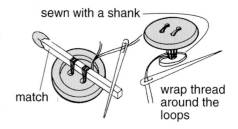

Fig 7.14 Attaching buttons

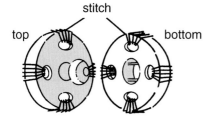

Fig 7.15 Attaching press-studs

Press-studs

- These are a simple and easy fastening for situations that are put under **little strain**.

- Although they are available in **coloured forms**, they are not usually used where they will be visible.

- They are extremely **difficult** to attach by machine and are generally **hand sewn.**

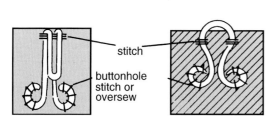

Fig 7.16 Attaching hooks and eyes

Hooks and eyes

- These are used in a similar way to **press-studs**.

- The **hook and bar** is a stronger version, which can be used on the **waistband** of skirts and trousers in place of buttons.

Velcro

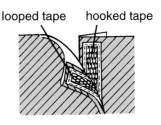

looped tape   hooked tape

Fig 7.17 Velcro tape

- This is a **two-part** tape used to close openings that will not be put under great strain, but require **easy access**.
- The effectiveness of the tape is severely **reduced** if the 'hook' part is **contaminated** with fluff and other fibres.

Zip fasteners

- Zips are the most useful form of **mechanical fastening** to use where the length of the opening needs to be closed and is likely to be put under **strain**. Even so, it is usually used in **conjunction with** another form of fastener, especially on trousers and skirts, to act as a back-up and to **protect the top teeth** of the zip.
- They can be fitted so that they are **visible or hidden**, with **closed or open** ends.
- They may also be fitted with **two-way openers** on outer garments, allowing easier access to inner layers.

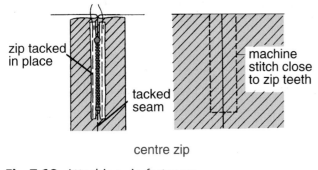

zip tacked in place
tacked seam
machine stitch close to zip teeth
centre zip

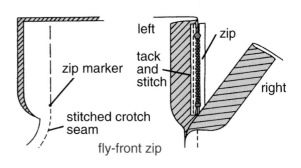

left
zip marker
tack and stitch
stitched crotch seam
fly-front zip
zip
right

Fig 7.18 Attaching zip fasteners

7.7 Fabric care

 LEARNING SUMMARY

After studying this section you should be able to:

- *discuss the requirements of washing fabrics*
- *understand the difference between ironing and pressing*
- *recognise standard care labels*

Washing, ironing and pressing

AQA
EDEXCEL
OCR
WJEC
NICCEA

Fabrics are washed to remove soiling and to help restore them to their original condition. Water, hot or cold, cannot do this efficiently so soap or detergent is added and the fabric agitated in some way – by hand or machine.

The **temperature, length and type** of wash cycle and the drying methods can all effect how well a fabric will wash.

 KEY POINT   It is important to take note of the fabric care label **and refer to the instructions on the detergent** packet **and the washing machine controls.**

Soap

- Soap is mainly made from **vegetable oils** and **caustic soda**.
- Soaps can be coloured and perfumed.

Care should be taken when using some detergents as they can cause allergic reactions in some people.

Detergent

- Made **synthetically**, detergents contain molecules that bond with the dirt particles, which are then removed by the agitation of washing machines.
- **Biological powders** contain **enzymes** that help digest the stains.
- Other **additives** can enhance the **brightness** of the fabrics after washing.

Ironing

Ironing **removes the creases** that washing and drying put in some fabrics. The **temperature** that the iron is set at is critical and so the instructions on the iron and the fabric care label must be followed.

Pressing

This is the process of deliberately **forming creases** in a fabric such as the creases in trousers or the pleats in a skirt.

Fabric care labels

AQA
EDEXCEL
OCR
WJEC
NICCEA

Flammability labels are statutory for nightwear and upholstered furniture.

Most fabric products will have a label attached, which must indicate:

- the **fibre content** of the fabric and their relative proportions
- **flammability** information, including relevant washing instructions

Optional information includes place of origin, sizes and care instructions – usually in the form of **symbols**.

> **KEY POINT** Many companies will wish their product to carry the kitemark – indicating that it conforms to certain standards issued by the British Standards Assurance Services.

Symbol	Instructions	Symbol	Instructions
	Machine code – this code, with a number in degrees inside, will indicate precisely the temperature of both machine and hand washing and the type of cycle to be used	⊠	Do not tumble dry
	Hand wash only	||||	Drip dry
	Do not machine or hand wash	 ⌂	Hang dry
Ⓟ	May be dry cleaned – other markings will indicate the process to be used	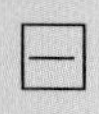 ⊟	Dry flat
⊗	Not suitable for dry cleaning	iron (cool)	Cool iron
△	Bleach (chlorine bleach) may be used	 iron (warm)	Warm iron
⊿	Do not bleach	 iron (hot)	Hot iron
⬚	May be tumble dried	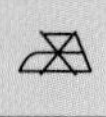 iron (crossed)	Do not iron

Fig 7.19 Symbols used in fabric care labels

Examination guidance and practice

Question formats

Types of question

Multiple choice

These are used rarely, but may appear in Foundation papers. The method is to discount those you know to be incorrect and think carefully about the rest.

- Other forms of this type of question are 'odd one out' and 'rank order'. Once again, think carefully, read what you have to do, and don't rush into ticking the most obvious answer.

Short answer

Commonly used in Foundation papers. The marks available and the space provided – usually on a pre-printed solid or dotted line – will give you an idea of the amount you need to include.

- Where possible, **use sentences** but be specific, e.g. 'coping saw', not just 'saw'.

Structured answer

> This is the most common form of question.

These usually start with a **stem** – a statement providing the context for the whole question, which will be in a number of parts.

- The subject may be new to you, unless it has been included in pre-examination resource/preparation sheets. It will have been chosen to allow you to demonstrate understanding and knowledge by reference to a **similar context** you are more familiar with.
- Sections of the question often get harder as they progress and will develop information given in previous parts. It is important you keep the **context of the stem** in mind at all times.

Open-ended

These are rarely used and are likely to appear only in Higher papers.

- An **extended answer** is required which you have to develop yourself. It is important to plan your answer and include all relevant points.

Resource/preparation sheets

Some boards issue these sheets **prior to the examination** in some specifications.

- They allow you to **focus your revision**, but it is unlikely that you will be allowed to take any prepared work into the examination.
- The information provided could include:
  - the **context** for the questions
  - some relevant **images or stimulus** material
  - hints as to the **materials and/or processes** that may be included

Key words used in questions

Annotate	• The addition of explanatory notes to enhance a drawing or sketch • Notes must be more than just labels
Choose/select	• Usually involves the selection of a given number of items from a list provided, or from a list you generated in an earlier part of the question, and commenting further in some way • E.g. 'Choose TWO' of the features you have listed and 'explain' why they are important
Comment	• Requires a brief analysis of a given situation • It is useful to back up statements with examples • Full sentences may not be necessary
Compare/contrast	• Both sides of an argument require discussion • Although you are likely to have to favour one side, marks will be given for your knowledge of both sides and the clarity of your argument for the decision you have made
Describe	• You must give an account in words, using sentences • Make sure that your answer is not just a list; marks are awarded for the quality of the description as well as the content
Discuss	• Requires an essay type response with the answer structured carefully and a conclusion arrived at • You may need to 'compare, justify and evaluate'
Draw/sketch	• Unless in a Graphic products examination, where these terms have specific meanings (see Chapter 4), you must provide a visual, non-verbal response • You are not marked on your 'drawing ability', but your sketch must clearly illustrate the points • Be careful to respond appropriately if the question includes 'labelled' or 'annotated'
Evaluate	• You are required to make a judgement from an extended analysis • The criteria on which to base this judgement may have already been generated in early parts of the question. If not, include the reasons for the conclusions you have reached
Explain	• Demonstrate an understanding of the point in question; it is not enough just to describe
Explore	• Use a variety of approaches – especially in design-based questions
Give	• Effectively the same as 'name' and 'suggest' • You can state your answer
Identify	• This is more than a list or selection • Produce a structured response in sentences
Justify	• Requires a well-argued reason why a 'selection' or 'judgement' has been made • Marks are awarded for the quality of reasoning as well as the content
Label	• One or two words should be added to drawings and sketches for further clarity
List	• Simply make a list
Name	• Similar to 'list', but often requires a fuller answer
Outline	• Used to show your knowledge of a complex issue in a broad way • Requires a less detailed response than 'describe'
Produce	• An instruction for you to do something • Generally involves drawing or writing after analysing some given information
Suggest	• Similar to 'name', but is often linked to a specific context
Using	• An instruction to clarify the way in which you should respond to a question • E.g. 'Using sketches, show how ...' • Marks will be lost if this instruction is not followed
Write	• Indicates that a written response is required and that sketches, tables, etc will gain marks only if they enhance the quality of the written aspects of your answer

Specimen questions and model answers

The Examiner will not try to test everything, as this would be impossible and unreasonable. The examination questions and the coursework criteria are carefully structured to allow you to display your **knowledge and understanding**, and it is up to you to prepare yourself effectively to achieve the best possible grade.

In the examination it is common practice for examiners to **group questions** based on a **common theme**. In some cases, the whole paper will be set in this manner, and a **resource sheet** will be issued prior to the examination.

 It is extremely important that your answers always refer to this theme, e.g. specifications noted in one answer are then carried out in the answers dealing with ideas.

The following 'examination paper' is made up of specimen questions and model answers from all five themes covered in the previous chapters. You are advised to look through them all as they cover the range of types of question that you could encounter.

 The following answers are only suggestions for how you could answer. In most cases there will be many other suitable responses.

Sample GCSE questions

Graphic products

1 A manufacturer of toiletries wants to introduce a new range of children's shampoo.

(a) Two specifications for the design of the container are:

- It should be suitable for use by 5–12 year olds of both sexes.
- It should be made from polythene.

Give THREE more specifications for the design of the container. **[3]**

- *The shape should be interesting to children.*
- *The top should control the rate of flow of the shampoo.*
- *It should be easy to hold when wet.* ◄─────

(b) State the reasons for the two specifications given in part (a), and the three you have given. **[5]**

- *So that all children in the target group will want the same product.*
- *For safety reasons - so that it would not break if dropped by the child.* ◄─────
- *To help children enjoy the experience of washing their hair.*
- *In order to avoid wasting the shampoo if it is dropped or the lid is left off.*
- *Younger children tend to play with such things in the bath.*

Food technology

2 A food manufacturer wants to introduce a new range of products for school cafeterias. The producer has carried out a survey of first school children and has identified the following specifications for a biscuit:

- large enough to be bought singly
- appealing to children because of the shape
- decorated

> *Make sure you **don't repeat** any specifications – especially the given ones!*
>
> *Use **sentences**, not one-word answers, when answering this type of question.*
>
> *List three **significant** specifications that relate **directly** to the design of the container.*
>
> *1 mark for each of three suitable specifications.*

> ***Avoid repetition** within the answers, e.g. referring to safety more than once.*
>
> *Simply **list** your reasons under **separate bullet points**, making sure that the reasons you give are **consistent with the theme** of the question.*
>
> *1 mark for each reason that correctly justifies the specification referred to.*

Sample GCSE questions

(a) Produce, using annotated sketches, TWO possible ideas for the shape of the biscuit. **[2 × 5]**

You could produce two sketches like this:

Football shape

80 mm diameter

MUFC

Iced topping

* Different teams could be done.

Hundreds and thousands

Iced topping

Ice cream 100 mm long

Smarties

*Present your sketches in a **3-D form**, e.g. isometric, **including dimensions**.*

*Draw them **boldly** and with **confidence**.*

*Be careful **not to spend too much time** on this type of question.*

*It is important that you **practise working at speed** as part of your revision programme.*

You should produce two sketches that:
- *are **significantly different** in concept*
- *show the **quality** of your sketching given the time available*
- *obey the **specifications** given in the stem of the question*

Marks for each sketch:
- *1–3 for the quality of the idea and the sketching*
- *1 for covering the specifications*
- *1 for relevant annotation*

(b) The biscuit is to be made using large scale batch production methods. Produce a simple flow chart to show FOUR main stages of the production, including ONE quality control check (feedback). **[6]**

Start

Select ingredients

Combine and knead

Roll and shape/cut

Check weight

Adjust quantities

Cook

Stop

*It is not necessary to use the **terminators** in this question, unless you have the time.*

*Concentrate on **selecting four distinct stages** in the **correct order** and an element of **feedback/quality control**.*

*Understand the use of at least these basic **flow chart symbols**.*

Marks:
1 for 2 suitable stages
2 for 3 suitable stages
3 for 4 suitable stages
1–2 for correct use of feedback
1 for the correct use of symbols

Sample GCSE questions

Resistant materials technology

3 A manufacturer has decided to make a prototype of a mobile kitchen storage unit in pine.

(a) State FOUR reasons for making a prototype. **[4 × 1]**

- *To check the design for such things as structural strength and stability.*
- *To check production methods and establish quality assurance procedures.* ← *1 mark for each suitable reason*
- *To check production costs.*
- *To carry out market research with potential customers.*

(b) Name a suitable construction joint that could be used for joining the runners to the uprights. **[1]**

- *Mortise and tenon* ← *Knock-down fittings should **not** be named. 1 mark for a correctly named joint*

(c) Name FIVE pieces of equipment used for marking out the joint. **[5]**

- *Mortise gauge*
- *Try-square* ← *Make sure you know the **correct name** for pieces of equipment.*
- *Steel rule* *1 mark for each of five*
- *Marking knife* *correct pieces of*
- *Pencil* *equipment*

(d) List THREE rules for marking out, giving the reason for each. **[3 × 2]**

- *Always work from a true or prepared edge – in order to ensure accuracy.*
- *Only make permanent marks that can be removed easily, e.g. by planing – so as not to spoil the finish of the piece.* ← ***Link** each reason with the relevant rule. 1 mark for an appropriate rule 1 mark for a correct reason*
- *Mark the waste side of any line to be cut – in order to avoid cutting on the wrong side.*

(e) In most cases, machinery would be used at some stage in making the joint.

(i) Name a piece of machinery that you could use in manufacturing the joint. **[1]**

- *Mortiser* ← *1 mark for a correct machine*

(ii) List TWO safety precautions you would take following a risk assessment. **[2]**

- *Isolate machine before setting up the chisel.* ← *Remember, 'wear eye protection' can often gain a mark here! 1 mark for each appropriate precaution*
- *The working piece should be securely clamped.*

Sample GCSE questions

Textiles technology

4 A manufacturer wishes to develop a range of products based on woodland animals. The first item to be developed is a backpack. The backpack is to be produced using batch production.

(a) What is the advantage of repetitive batch production?   **[2]**

- *Because the numbers of each design would not require continuous production, the basic equipment can be set up to produce a set number (say 500), and then move onto another design. All jigs, patterns and programmes can then be stored for future use if sales generate the need.*

> *Include as much* **technical information** *as you can.*
>
> *1–2 marks for a correct description of the advantages*

(b) What is the main benefit of using CAM (computer-aided manufacture) in the production of the backpack?   **[2]**

- *The main benefit is that consistent manufacture is achieved in terms of measurements.*

> *Be* **specific** *– don't just say 'better quality'.*
>
> *1–2 marks for a correct reason*

(c) Describe a quality control process that may take place during the manufacture.   **[2]**

- *The seams would be checked for strength on a regular sample of the backpacks.*

> *Make sure you know the difference between* **'quality control'** *and* **'quality assurance'**.
>
> *1–2 marks for a correct process*

(d) State one way in which the consumer can check on the quality of a product such as this.   **[2]**
- *The 'kite mark' or other seal of approval should be attached.*

> *Make yourself aware of the various safety labels.*
>
> *1–2 marks for a correct method*

Systems and control technology – electronic products

5 A CMOS 8 pin Integrated Circuit is shown below.

Sample GCSE questions

(a) Insert the pin numbers. **[1]**

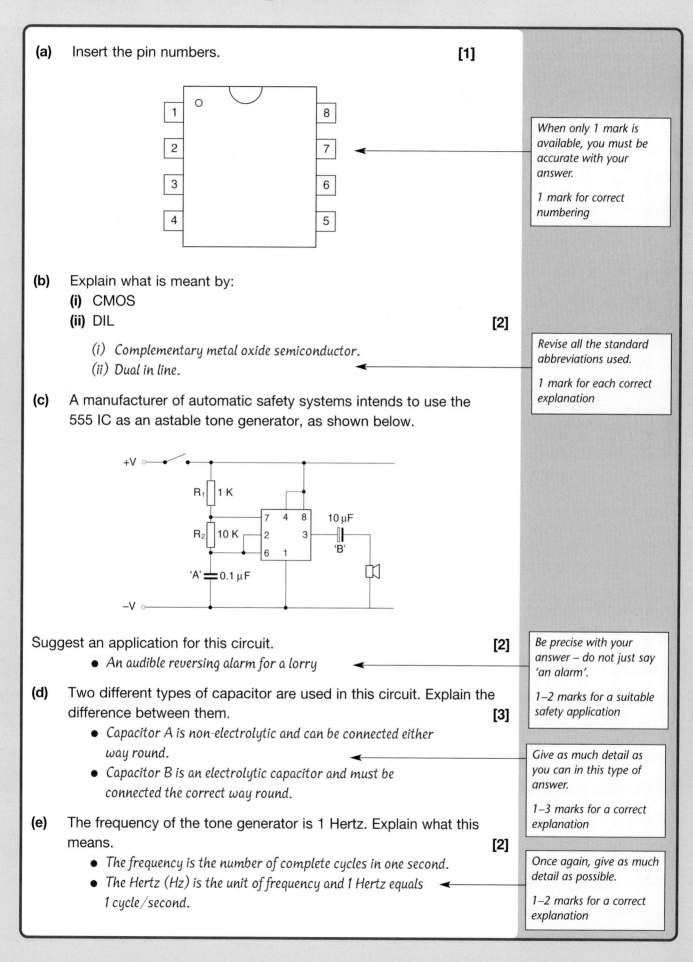

> When only 1 mark is available, you must be accurate with your answer.
>
> 1 mark for correct numbering

(b) Explain what is meant by:
(i) CMOS
(ii) DIL **[2]**

(i) Complementary metal oxide semiconductor.
(ii) Dual in line.

> Revise all the standard abbreviations used.
>
> 1 mark for each correct explanation

(c) A manufacturer of automatic safety systems intends to use the 555 IC as an astable tone generator, as shown below.

Suggest an application for this circuit. **[2]**
• *An audible reversing alarm for a lorry*

> Be precise with your answer – do not just say 'an alarm'.
>
> 1–2 marks for a suitable safety application

(d) Two different types of capacitor are used in this circuit. Explain the difference between them. **[3]**
• *Capacitor A is non-electrolytic and can be connected either way round.*
• *Capacitor B is an electrolytic capacitor and must be connected the correct way round.*

> Give as much detail as you can in this type of answer.
>
> 1–3 marks for a correct explanation

(e) The frequency of the tone generator is 1 Hertz. Explain what this means. **[2]**
• *The frequency is the number of complete cycles in one second.*
• *The Hertz (Hz) is the unit of frequency and 1 Hertz equals 1 cycle/second.*

> Once again, give as much detail as possible.
>
> 1–2 marks for a correct explanation

Sample GCSE questions

(f) Using the component values given in part (c), calculate the operating frequency using:

$$f = \frac{1.44}{(R_1 + 2R_2)C} \text{ Hz}$$

[2]

$$\bullet\ f = \frac{1.44}{(0.001 + 0.02)0.1} = \frac{1.44}{0.021 \times 0.1} = 685\,Hz$$

> *Take care to get the decimal points in the correct place!*
>
> *Marks:*
> *1 for using the correct values*
> *1 for the correct answer*

(g) It has been decided that, to operate in different working conditions, the tone generator should be adjustable. Redesign the circuit in part (c) so that this is possible. **[4]**

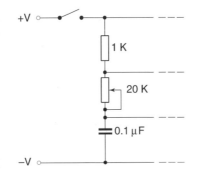

> *1–4 marks for the inclusion of a variable resistor of a suitable value, in the correct location R_2 for full marks*

Index

Index